HUMAN MOTION ANALYSIS

OM MISHRA

Copyright © [2023]

Author: Om Mishra

Title:. Human Motion Analysis

All rights reserved. No part of this publication may be reproduced or transmitted in any form or by any means, electronic or mechanical, including photocopying, recording, or any information storage and retrieval system, without prior written permission from the author.

This book is a self-published work by the author Om Mishra

ISBN:

For permission to reproduce any of the material in this book.

TABLE OF CONTENTS

List of Figures
List of tables.
1. Introduction 1
   1.1 Literature Review 3
   1.2 Preprocessing Methodologies 3
      1.2.1 Background Subtraction Based Preprocessing 3
      1.2.2 Optical Flow Based Preprocessing 5
      1.2.3 Filter Response Based Preprocessing 5
   1.3 Modeling of Human Action 6
      1.3.1 2-D Temaplate Matching Techniques 6
      1.3.2 Space-Time Volume Based Technique 6
      1.3.3 Spatio-Temporal Interest Points Based Technique 7
      1.3.4 Deep Learning Based Technique 7
      1.3.5 Silhouette Analysis-Based Methodologies 8
   1.4 Anomaly detection in Human Motion 11
   1.5 Classification 12
   1.6 Challenges of Human Motion Analysis 13
   1.7 Research Gap 17
   1.8 Research Objectives 18
   1.9 Research Contribution 19
2. Feature descriptor based on Selective Finite Element AnalysiS 21
   2.1 Motivation for the Proposed Framework 21
   2.2 Methodology 21
      2.2.1 Silhouette Extraction 22
      2.2.1 Discretization and Shape Function Representation 23
      2.2.3 Representation of Feature Vector 26
      2.2.4 Dimension Reduction and Classification 28
   2.3 Algorithm 28
   2.4 Experimental Setup 29
      2.4.1 Datasets Used 30
      2.4.2 Parameter Settings 31
      2.4.3 Results Discussion 33
   2.5 Run Time Analysis 39
   2.6 Chapter Conclusion 40
3. Modified Bag-of-Visual-Words based descriptor 41
   3.1 Introduction 41

3.2 Proposed Methodology 42
    3.2.1 Geometrical Description of a Cluster 44
    3.2.2 Contextual Distance among the points of a Cluster 45
    3.2.3 Formation of the Directed Graph of a Cluster on the Basis of Contextual Distance ... 46
    3.2.4 Laplacian of the Directed Graph of a Cluster 48
3.3 Parameter Setting 49
3.4 Results 50
    3.4.1 Analysis and Discussion on KTH 50
    3.4.2 Analysis and Discussion on Ballet Dataset 52
    3.4.3 Analysis and Discussion on IXMAS dataset 54
3.5 Chapter Conclusion 56
4. Descriptor based on Fuzzy lattices using Schrödinger wave equation **57**
4.1 Methodology 57
    4.1.1 Formation of Fuzzy Lattices 58
    4.1.2 Dimension Reduction and Classification 62
4.2 Datasets Used 64
4.3 Parameter Sensitivity Test 66
4.4 Performance Evaluation Methodology 67
    4.4.1 Performance evaluation for UMN Dataset 68
    4.4.2 Performance evaluation for UCSD Dataset 69
    4.4.3 Performance evaluation for UCF Dataset 71
    4.4.4 Performance evaluation for YouTube video and New MOT 17 dataset 73
4.5 Real-Time Applicability 75
    4.5.1 Run-time Analysis 75
4.6 Chapter Conclusion 77
5. Conclusion and Future Scope **78**

TABLE OF FIGURES

Figure 1.1 Background subtracted image for action 'walk' 4
Figure 1.2 Optical flow based processed image 5
Figure 1.3 Challenges of action recognition 13
Figure 2.1 Workflow diagram of the proposed method 22
Figure 2.2 Walk and the extracted silhouette 22
Figure 2.3 Segmented discretized silhouette 23
Figure 2.4 Displacement of a mesh (triangle) of the human body and its mesh element 23
Figure 2.5 Deformed triangle as node 1 is displaced 23
Figure 2.6 Division of the triangle into 3 parts 23
Figure 2.7 Weizmann dataset 30
Figure 2.8 KTH dataset 30
Figure 2.9 Ballet dataset 31
Figure 2.10 IXMAS dataset (5 cameras) 31
Figure 2.11 Parameter setting for a. Number of nodes; b. Number of finite elements; c. Young's Modulus; d. Feature dimension through PCA 32
Figure 2.12 Comparison of the proposed method with similar methods for four datasets ... Weizmann, KTH, Ballet, and IXMAS 36
Figure 3.1 Detail process flow diagram of the proposed methodology 44
Figure 3.2 Geometrical description of a cluster and contextual distance among the cluster points 45
Figure 3.3 A cluster set Ic of m points where A=$ic0$, B= $ic1$ and λ (A, B) = ($ic0$, $ic1$) 47
Figure 3.4 Formation of N clusters and their Laplacian descriptors 49
Figure 3.5 Selection of codebook size 50
Figure 3.6 Selection of size of nearest neighbor 50
Figure 3.7 Confusion Matrix for actions in KTH dataset 50
Figure 3.8 Confusion Matrix for actions in Ballet dataset 53
Figure 3.9 Confusion Matrix for actions in IXMAS dataset 55
Figure 4.1 Block diagram of proposed method 57
Figure 4.2 Connected Pixels 59
Figure 4.3 Fuzzy lattices 59
Figure 4.4 Process flow diagram for the proposed technique 64
Figure 4.5 UMN dataset normal frame 68
Figure 4.6 UMN dataset abnormal frame 68
Figure 4.7 Result on UMN dataset (X axis represents frames and Y axis represents change in kinetic energy) 68
Figure 4.8 UCSD dataset normal frame 70
Figure 4.9 UCSD dataset abnormal frame 70
Figure 4.10 Result on UCSD dataset (X axis represents frames and Y axis represents change in kinetic energy) 70
Figure 4.11 UCF dataset normal frame 72
Figure 4.12 UCF dataset abnormal frame 72
Figure 4.13 Result on UCF dataset (X axis represents frames and Y axis represents change in kinetic energy) 72
Figure 4.14 Youtube dataset 74
Figure 4.15 MOT 17 datase 74
Figure 4.16 Result on Youtube video (X axis represents frames and Y axis represents change in kinetic energy) 74
Figure 4.17 Result on MOT 17 dataset (X axis represents frames and Y axis represents change in kinetic energy) 75

TABLE OF TABLES

Table 2.1. Confusion matrix for Weizmann Dataset (R-Running, W-Walking, J-Jumping, JJ-Jumping Jack, S-Skipping, JP-Jumping at a place, SJ-Side Jumping, B-Bending, W-Waving with one hand, WB-Waving with both hands) 34

Table 2.2. Confusion matrix for KTH Dataset (A- Applauding, W- Waving, B- Boxing, WK- Walking, J-Jogging, R- Running) 34

Table 2.3. Confusion matrix for Ballet LR- Left to right-Hand Opening, RL- Right to left-Hand Opening, J-Jumping, H- Hopping, S- Swinging leg, ST- Standing, T- Turning) 34

Table 2.4. Confusion matrix for IXMAS Dataset (W- Walking, WA- Waving, P- Punching, K- Kicking, T-Throwing, P- Pointing, PU- Picking Up, G- Getting Up, S- Sitting Down, TA-Turning Around, F-Folding arms, C-Checking Watch, SH-Scratching Head)... 35

Table 2.5 Comparison of the proposed method with similar methods on Weizmann dataset 37

Table 2.6 Comparison of the proposed method with similar methods on KTH dataset............................. 38

Table 2.7 Comparison of the proposed method with similar methods on Ballet dataset.......................... 38

Table 2.8 Comparison of the proposed method with similar methods on IXMAS dataset......................... 38

Table 3.1 Comparison among other state-of-the-art methods for KTH.. 52

Table 3.2 Comparison among similar state-of-the-art methods.. 52

Table 3.3 Illustration of EER and Accuracy for various approaches on Ballet dataset 54

Table 3.4 Accuracy for various approaches from 5 different cameras 55

Table 4.1 EER and AUC as anomaly detection parameters on UMN dataset 69

Table 4.2 Confusion Matrix of proposed method and other existing methods local dataset where N-normal and A-abnormal.. 69

Table 4.3 EER and AUC as anomaly detection parameters on UCSD dataset 71

Table 4.4 Confusion Matrix of proposed method and other existing methods local dataset where N-normal and A-abnormal.. 71

Table 4.5 Equal error rate for anomaly detection comparison for Local dataset....................................... 73

Table 4.6 Confusion Matrix of proposed method and other existing methods local dataset where N-normal and A-abnormal.. 73

Table 4.7 Run-time analysis of different methods... 76

Table 4.8 Comparison of proposed method with other dimension reduction technique 76

1. INTRODUCTION

In this chapter, we discussed about human motion analysis, its application and challenges. The most important application of the human motion analysis is Human action recognition. In our research, we recognized human action by analyzing human motion. We discussed various methodologies to extract the action features. We also discussed the challenges that affect the rate of human action recognition. At the end of this chapter, we discussed the research gaps and the objective of our research. Later on, we discussed our contribution to achieve these research objectives.

Over the last few decades, it has been observed that computers have transformed human life in almost every possible aspect. Along with the latest transformations, video data has become easily accessible and dominant in the present time. Every new reform has enabled hardware devices like mobile phones, tablets, digital cameras to create, store and share videos. The increasing number of accessible videos has also created the need to understand them. This idea has led to an extensive study of the analysis of human motion in the video in the field of computer vision and image processing. Human motion analysis has its major application in the field of medical, sports and security. In this research we have used human motion analysis for human action recognition in a video.

The analysis of human motion is used to recognize the human action or activity in the video. Most of the researchers used 'action' for single person's motion like 'Walking', 'Running', 'Jogging' etc. The term 'activity' is used by the researchers for the motion pattern of multiple persons. These activities may contain normal or abnormal activities. The motivational applications of the human action recognition are discussed below:

Recently, we have seen that analysis of crowd activities is very important specifically in the part of South Asia, due to a large population of countries such as India, China, and Pakistan. It is very tough to monitor crowd activities of busy places. In India, so many incidents of stampede occurred in different parts of the country. Terrorism is also a major problem in this part of the world. To monitor these kinds of activities, there is an increasing demand for automatic visual surveillance systems. In this case video captured by CCTV camera is analyzed to detect any abnormality in the scene.

Human motion analysis is also used for behavioral biometric. We can recognize the person as he or she performs the actions. The other biometric such as iris recognition, face recognition, fingerprint recognition etc. requires the physical contact as well as the concern of the persons whose biometric is to be taken. But we can recognize the person from distance and without taking his/her concern through the analyzing the human motion in the video. This behavioral biometric is also known as the gait. These gaits are already used to identify the person for the conviction in the judicial court.

As we all know that due to increasing number of cameras, videos and the pictures play major role in our daily life. We share these videos through emails and social media etc. To extract the content of this large amount of video data has become very important. Human motion analysis can be used to extract the important information from the large raw videos. Human motion analysis can also be used in the human computer interaction as well as in gaming industries.

1.1 Literature Review

In general, system of human motion analysis requires the following steps:

1. Input video
2. Preprocessing
3. Feature descriptors to model the action
4. Classification for action recognition

In the analysis of human motion, we focused on the action recognition in the videos. The action video has large number of raw information. To recognize a human action there is a need to extract the low level feature. Basically, these features include preprocessing. The following methodologies have been prominently used for preprocessing:

1.2 Preprocessing Methodologies

1.2.1 Background Subtraction Based Preprocessing

These methodologies are used to segment background and foreground. Basically, foreground is the moving part whose action is to be recognized. Figure 1.1 shows some background subtracted frames from the video of action 'walk'. Some popular methods for background subtraction methods are: methods based on frame differencing, methods based on Kernel estimation, methods based on Gaussian Mixture Model, methods based on thresh-holding techniques, method based on PCA etc.

Figure 1.1 Background subtracted image for action 'walk'

In frame differencing methodologies, the consecutive frames are subtracted with each other. But the problem with these methods is that it cannot handle dynamic background. The kernel estimation method [3] used novel background subtraction method. This method also deals with the problem of background cluttering and shadow. This methodology estimates the probability of the intensity value of each pixel in the image sequence. If there is any change in the intensity of the pixel due to motion, this method immediately shows the change. The codebook based background subtraction methodology in [4] modeled the background by quantizing the value of each pixel of the image into codebook. This method dealt the problem of changing background and illumination change. The methodology based on Gaussian Mixture model is used to model dynamic background and also to solve the problem that occur due to the shadow. Methodology in [5] used the mixture of Gaussian to model each pixel of the image and on the basis of mean and variance of the Gaussian Mixture the background color is selected. The value of the pixel which does not relate to the background distribution is considered to be foreground. In [6] each pixel of the image is modeled by Gaussian covariance matrix. The methodology based on self organized neural network is used for background subtraction in [7]. In [8] random aggregation methodology is used where neighboring pixels are used to model the background. A thresh-holding technique based methodology in [9] used 3D-geometry for background subtraction. Methodology in [10] used advanced color based algorithm where color and depth are fused to extract the foreground. Methodology discussed in [11] used color and depth information based on Gaussian Mixture Model.

1.2.2 Optical Flow Based Preprocessing

Optical flow [12] gives the apparent motion of the pixel. The optical flow shows both velocity and direction of an image pixel. The optical flow based methodologies assumes that intensity values of the pixels do not change from one frame to other frame of the video. The problem with these methods is that it very sensitive to noises. Figure 1.2 shows optical flow based processed image.

Figure 1.2 Optical flow based processed image

1.2.3 Filter Response Based Preprocessing

Filter response based features are also used for preprocessing of the action video. The spatial and derivative of Gaussian are applied to the image frame on the temporal axis in [13]. They give motion information of the action videos. Gaussian derivative filters are also used to find out the spatio-temporal prominent points in moving object [14]. Methodology in [15] used Gaussian filters in the space and Gabor filter in time to preprocess the frames of the video. These methodologies used convolution that made the implementation easy.

1.3 Modeling of Human Action

1.3.1 2-D Temaplate Matching Techniques

The 2D template methods use 2D silhouettes for global representation in [16 and 17]. Methodology in [16] used motion energy image and motion history image to represent human action video globally. Motion energy image gives the information about location of the motion and motion history image gives the information about how the motion has been performed. Methodology in [17] used Radon transform to describe the action video. The advantage of Radon transform is that it is invariant to translation, rotation and noise.

1.3.2 Space-Time Volume Based Technique

Spatial-temporal volume is made by stacking silhouettes over a given grouping [18 and 19]. Methodologies in [18 and19] used two dimensional shapes to create space time volume. Space time volume based method contains both spatial as well as pose based information. This method deals with the problem of variation in view point. Action features may also represent motion information with the assistance of optical flow. It doesn't require background subtraction. An optical flow-based strategy where the movements of pixels have been contemplating is also used in global feature extraction in [20 and 21]. The disadvantage of this method is that it is very sensitive to noise because the motion descriptor can be corrupted due to noise that appeared in a dynamically changing background.

1.3.3 Spatio-Temporal Interest Points Based Technique

Local spatiotemporal descriptors are based on the bag-of-words model [22-24]. Methodology in [22] used the space time interest point in the action video. These interest points are represented as sparse space time features that are used to build the bag-of-feature. Methodology in [23] used the bag-of-feature technique where correllogram based features are used. In correllogram, co-occurance of the spatio temporal features are obtained. Methodology in [24] used the relationship among the spatio temporal points. The histogram of three dimensional gradient orientation is used as feature descriptor in [26]. Methodology in [27] proposed a new scale time invariant technique. The scale invariant feature is represented by Hessian matrix extracted from the scale-space theory. Methodology in [28] proposed an algorithm based on scale invariant features that are capable to detect those interest points which are distinctive due to the motion. This Feature descriptor is known as MoSIFT.

1.3.4 Deep Learning Based Technique

Furthermore, in recent years, a paradigm based on deep learning techniques is also very popular in the research community for human action recognition [29-31]. The approaches based on deep learning technique [32-34] are fully automated. Sabokrou and Fayyaz [32] used the 3D convolution to extract the features. Unlike the handcrafted approaches discussed above, the deep learning technique is fully automated. The efficiency of deep learning methods depends upon the design of the network. These methods need large datasets and parameters. On account of this, the complexity of the structure increases. Researchers are working on deep learning techniques to overcome this problem.

1.3.5 Silhouette Analysis-Based Methodologies

Silhouettes extracted from the frames contribute significantly to human action recognition. They are used to find out both global and local features in [35- 21]. Methodology in [35] used the average energy image of the action silhouette. Spatial distribution of gradients is then applied to describe the action. This gives the shape information of the action. Radon transform is used to get motion information. In the methodology of [36], the human silhouettes are divided into fixed number of cells and grids. The hybrid classifier (KNN, SVM and LDA) is then used to classify the actions. Methodology in [37] used the normalized silhouettes of action video to find the bag-of correlated-poses. These features are fused with features extracted from the extended motion history image. They also used new kernel codebook to remove the problem of quantization in the bag-of-feature model.

Methodology in [38] used 3D histogram of gradients to describe the action video. These features are concatenated to form a data volume. Methodology in [39] used both silhouettes and contours to model the action. Histogram of oriented gradient (HOG) of the silhouettes is used as the action descriptor whereas in contour based approach Fourier descriptor, chord length feature, centroid distance features and Cartesian coordinate are used as feature descriptor. Methodology in [40] used human depth silhouettes and human skeleton joints to represent action. Features like distance based on torso, distance based on key joints and directional angle are calculated from human skeleton joints. The HOG is applied on depth silhouettes to find out differential silhouette between two frames.

In the methodology of [41], silhouette extraction is done through graph based approach. The RGB depth action videos are used in [42]. This method used spatio-temporal kernel descriptor to represent the actions. In [43] a robust methodology for silhouette extraction is used based on Gaussian Mixture Model and Laplacian scheme.

In the methodology of [44] Gaussian Mixture Model is used for silhouette extraction. Motion history image and motion energy image are then formed from these silhouettes. Spatio-temporal interest points are then found out to describe the actions. Methodology in [45] used the sparse representation of the dictionary in the context of action recognition. They used distinctive spatio temporal features. Methodology in [46] used motion history volume to represent human actions. Fourier Transform is then used as feature descriptor. A new technique where the surrounding of the region of the subject is emphasized is used in [47]. This region is called negative space. First of all, the silhouette is extracted to represent the pose of the action, then negative space based descriptor is obtained to model the action.

Methodology in [48] proposed a new technique based on differential geometric trajectory cloud. In this technique Frenet-Serret Equation is used to characterize the local geometry of the human action. Bag-of-visual-word is used in [49] where frames of the video are considered as a 'word'. They improved the performance in terms of time by using reduced number of parameters. In the methodology of [50], the skeleton is first obtained. The reference joints are then selected in the skeleton using graph matching. Methodology in [51] used the new bag-of-visual-word based method to represent the action. They also used distinctive codebook generation to remove the recognition error. Each action is represented locally as set of XYT coordinate between the two interest points in [52]. The cuboid centered on the interest point is created to capture the shape information of the action. Methodology in [53] described the action video using three dimensional histogram oriented gradients. These descriptors formed the data volume of the action. Methodology in [54] proposed a technique where multiple cameras are installed to observe the activities from different views. The convex optimization is used to form a distributed matrix as action features. A statistical translation framework is used for modeling the action in [55]. The K mean clustering is used to form bag-of-visual-word to represent local action features. The key poses of human body represent an action effectively in [56 and 57]. The distance among the spatio temporal points are used in [56]. Multi-View key poses from the silhouette are extracted in [57]. The contour points on these silhouettes are found out. This method assumes no change in the temporal order of the pose sequences. The binarized silhouette is used to find out the trace transform to represent the global feature of action in [58]. This trace transform is the general form of the Radon transform.

Authors of [59] proposed a new methodology where two dimensional silhouettes are taken from different views. The features are represented by new multi-dimensionality reduction technique along the spatial and temporal axis. Bandelet transform is used to extract the geometrical feature from the silhouettes in [60]. These features give the information of the shape change in human body. To select the prominent features Adaboost algorithm is used. Methodology in [61] first divided the input action video into patches. These patches are divided in such a manner that they do not overlap each other. The Shanon's theory for entropy is used to find out the space time features. The silhouette-based analysis is also used in deep learning-based methodologies in [62, 63, and 64]. Methodology in [62] is used for recognizing continuous action in the video. The framework used hybrid architecture of Convolutional Neural Network and Hidden Markov Model. The Convolutional Neural Network is used to extract action features from the raw data. The Hidden Markov Model is used to find out the statistical dependency among the two action class. Methodology in [63] used the model that is bio-inspired. They used primary visual cortex and middle temporary cortex. Author of [63] proposed a sequential deep trajectory descriptor to represent long motion. CNN and RNN are then used. It captures static spatial features to represent long as well as short term motions.

1.4 Anomaly detection in Human Motion

The methods discussed above are mostly based on the detection and tracking of the subjects. But to analyze the anomaly occurring in the scene due to human motion, one very popular category is based on trajectory modeling [65-73]. The trajectory-based translation invariant feature is used in [65] to understand the videos. The trajectory of the agent is used to find out its behavior pattern in [66]. Authors have used motion pattern and size pattern of the subject as the features to detect the anomaly in the scene [67 and 68]. In [71], the likelihood of the trajectory of an object is calculated to understand the scene. A Bayesian approach is used for an anomaly in the scene in [72]. In the methodology of [73], statistical distance measures are used to find event detection in two videos. In these methods, moving targets are tracked and classified into different categories that make use of image features and image-sequence-based tracking results. The main drawback of these methods is their sensitivity to occlusion and tracking errors. Various authors have proposed alternative motion representations that avoid tracking. Spatial and temporal anomalies in the scene are modeled by the appearance and dynamics of the scene in methodology of [75 and 78]. In [74], anomalous events in the scene are detected by creating a number of local monitors. Local statistical data from different monitors are fused to make a decision about an anomaly in the scene. The space-time Markov Random Field (MRF) is used to detect an unusual event in the video in [76]. The features thus extracted are directly used to create models of activities. These techniques suffer a curse of dimensionality which makes them non-compatible for real-time implementation.

1.5 Classification

In this section we review some classification techniques that have been used in Human Action Recognition. The Support Vector Machine [102, 103 and 104] and Nearest Neighbor Classifier (NNC) [106, 107 and 108] are used by the researchers because they are simple and effective in action and anomaly recognition. Most of these methods used Euclidean distance between two features. In the case of low dimensionality Euclidean distance is effective but it is not effective for high dimensionality problem. Thus, distances like Weighted Euclidean Distance [109], Mahalanobis Distance [110] and Chamfer Distance [111] are used for high dimensionality problem by the researchers. The main problem with these distances is that they assume the distribution of training and testing data do not change. They also require large training data to cover all variation in the action pattern.

1.6 Challenges of Human Motion Analysis

There are some significant challenges faced by the researchers in human action recognition. We have identified and classified some of the important challenges in human action recognition. These challenges are cluttered background, camera motion, variation in view point, occlusion and variation in execution rate as shown in Figure 1.3.

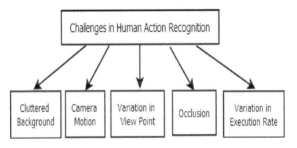

Figure 1.3 Challenges of action recognition

Cluttered Background: Change in the background or background cluttering distracts the concentration of point that should originally be the action to be recognized. The disadvantage of most of the methods was that they were not good enough for complex backgrounds. Background subtraction is also not easy when background is changing. A method based on background modeling such as Gaussian mixture models (GMM) [5] is used to model the dynamic backgrounds. These methods are also capable to deal with the problem due to shadow. The difficulty with these models is that it is hard to model background for a long duration. To deal with the problem of background modeling some researchers have suggested methodologies based on volumetric spatiotemporal features

[18-19]. These methods are robust to background change because they do not require background subtraction and these features are robust to the noise and change in illumination as well. Some other very popular features that are robust to the background changes are bag of words based codebook methods [22-24]. Frames of the action videos contain key poses of the actions. To avoid background variation some researchers proposed the key pose based approach [56-58]. The main problem with this approach was that they only rely on the frames having key poses that lead to loss of the information of the video and affects the recognition rate also. Thus, for background cluttering, model based on GMM and spatio-temporal features performed well as compared to other methods. The disadvantage of GMM is that it is very time consuming.

Movement of Camera: In general most of the methods for action recognition assume that the camera is static and placed on a particular location. But if camera is moving then it creates a big challenge in action recognition because of change in motion. To deal with motion of camera researchers have suggested methods like optical flow [20-21] and local space time features which are not affected by camera motion in terms of interest point. The disadvantage of this method is that they are very sensitive to noise. Methods based on codebook are also used to deal the camera motion. In codebook method, a dictionary is created based on the local features and features are clustered in the code word. The clustering incorporates the quantization error problem that affects the recognition of the action. To summarize, codebook method give good results but their prerequisite is that they require proper camera calibration.

Variation in Execution Rate: The speed of the action being performed also contributes to degradation in the recognition rate. Human body is very flexible and thus may not perform every action at same speed each time. Even two individual cannot perform action at same speed. Someone may walk slowly someone may fast. Researchers have suggested dynamic time warping based approach [114-115]. In dynamic time warping approach whole video of the action is taken into account. Input video and the template are being compared at the same time scale to deal with the change in execution rate. Method which do not capture the motion features such as histogram based methods are also used for variation in execution rate. Methods used codebook based method using histogram approach to eliminate the effect due to the variation in execution rate. But as discussed above, codebook method suffers from quantization error. Some researchers proposed probabilistic approach as the solution of this problem. Probabilistic methods such as Hidden Markov Model, Bays network [116-117] are used to model action. They only rely on the probability of the state change and are independent of execution rate of the action. The drawback of these methods is that they require huge training dataset to model the action effectively. Among all the discussed methods for varying execution rate, dynamic time warping methodology seems a better method as compared to others.

Variation in View Point: Variation in view point is the most challenging factor in action recognition. In most of the existing methodology camera is fixed. But methodologies which are made for the fixed camera location may not be applicable for other view. The appearance of the object changes if camera location is changed. It leads to the change in the motion pattern for same action video. This affects the recognition rate badly. Method based on Radon transform [118] is the most popular view invariant features. In radon transform 2-D object extracted from the video frames is projected onto radon space. But the problem of this method is that it requires sophisticated background subtraction to detect the object. Also in Radon transform we calculate the line integral. This indicates that Radon Transform only captures low frequency component leaving the high frequency components. Fourier transform [119] and cylindrical coordinate system [120] are also used as view invariant features. Self similarity and dissimilarity are also used as view invariant features. Researchers used methods based on multi view multiple cameras to overcome the challenge [121]. In this approach multiple cameras are installed at different angle. These angles are evenly spaced. Methodology which is applicable for one camera is applied for others also. Classifiers are used for different camera and at the end all classifiers are fused. This is a good solution, but it increases the complexity in terms of both computation and cost.

Occlusion: Occlusion is also a very important issue that can degrade the action recognition rate but in general most of the methods assume that the object is clearly visible while performing action in action video. But this may not be true for real time scenarios. Occlusion is of two types. One is when an object is occluded with other object and second is self occlusion where some body part of the object is occluded by other body part of the object. Probabilistic approach based methods like Hidden Markov Model [116], Bayesian network [117] are used to avoid occlusion. In these methods human body is separated into different parts. Each part of the body is considered as the state and change of state is defined by the probability value. These methods require the tracking of the body part in subsequent frame. These methods need background subtraction to be done very carefully. Spatio-temporal volumes are also used to avoid occlusion. This representation is more robust to both types of occlusion. Features represented by space time volume involve volumetric representation of the object in the entire video. The volumetric representation of the body will give enough representation from the frames as object is not occluded throughout the video. To avoid self occlusion completely, methods based on multiple camera view gives better result, but it increases the computational time and cost. Spatio-temporal volumetric representation gives the best recognition as compared to others.

1.7 Research Gap

On the basis of literature review we may divide the modeling of human motion analysis into two categories:

1. Global Feature Descriptors
2. Local Feature Descriptors

Researchers modeled action features globally as well as locally. Global features descriptors in [16, 17, 18, 19, 20 and 21] rely on the localization of individual whose action is to be perceived. Localization is done through background subtraction or human tracking. Primary inconveniences of global features portrayal are that it unpleasantly depends on precise localization and background subtraction, so it is sensitive to a viewpoint and individual appearance. Likewise, it cannot give motion information of an activity that makes it sub-par compared to similar kinds of activities like jogging and running. Local feature descriptors in [22, 23, 24, 25, 26, 27 and 28] are utilized more frequently in recent times. These descriptors give motion information. The disadvantage of local feature descriptor is that they can't give basic/shape information. Its after-effect is that it can give the same component descriptor for various activity classes. To avoid these issues hybrid feature descriptors are used in [37, 38, 39, 40 and 41] to capture both shape and motion information of the action. They suffer large feature dimensionality problem that is a big hurdle for real time application.

1.8 Research Objectives

The main objective of my research is to analyze the issues of the existing methodologies and to propose new framework which can overcome the problems related to Global feature descriptors and Local feature descriptors. We also aimed to propose new methodology for anomaly detection in the crowd activity. As discussed above, the existing methodologies suffer the problem of dimensionality. Thus, one of our research objectives is to deal the problem of dimensionality. To summarize, the research objectives are as follows:

1) To propose a new feature descriptor which can represent shape as well as motion features.
2) To detect anomaly in the video.
3) To propose a new methodology for real time application.

1.9 Research Contribution

During our research, I worked to solve the issues related to existing action descriptors for action recognition. As discussed above, there are two categories of descriptors, one is Global Feature Descriptor and other is Local Feature Descriptors. The problem with the global feature descriptors is that they only give shape information of the human body. They do not give motion information of the action. This makes the accurate action recognition of similar type of actions where motion is dominant difficult. The local features deal the motion information but they may not give shape information. This creates problems in the accuracy of the action recognition. Some researchers have proposed hybrid descriptors that used both global as well as local descriptors. But they suffer with high features dimensionality problems. Our contribution to solve these issues is as follows:

1. We proposed a new local feature descriptor using Finite Element analysis, which is solely capable of representing shape as well as motion information of the action.
2. To overcome the issue such as variation in view point, background cluttering and occlusion, the bag-of-visual-word based methodology is very popular. But the main problem with these methodologies is that while making cluster they do not retain the structure of the cluster. In our research I proposed a modified bag-of-visual word based method which takes care of structure of the clusters.
3. We also proposed a new methodology for anomaly detection in the crowd activity. We used fuzzy lattices and these fuzzy lattices are described by the

kinetic energy of the fuzzy lattices using Schrödinger wave equation. This methodology has reduced feature dimension that makes it real time applicable.

2. FEATURE DESCRIPTOR BASED ON SELECTIVE FINITE ELEMENT ANALYSIS

2.1 Motivation for the Proposed Framework

This chapter introduces a new feature descriptor based on Finite Element Analysis (FEA) [123 and 124]. FEA has been used as a very powerful technique for the structural analysis of the system. In the FEA technique, the structure is converted into a finite number of elements. Wherever any deformation occurs in a body/structure these finite elements also get displaced from their previous position and the stiffness matrix of these elements shows how stiff the body/structure is against this deformation. This gives accurate and precise information about the structural deformation of the body. Similarly, when a person performs an action, his body gets deformed in different patterns. This motivates us to apply the concept of FEA on the silhouettes extracted from the action video. The proposed method offers a new local features descriptor that is solely capable of representing shape as well as motion features of the silhouette.

2.2 Methodology

Figure 2.1 describes the proposed methodology. The human silhouette is extracted from the frames of the action video. Then we discretized the silhouette of the human body into several finite elements (triangle faces). Then complete stiffness matrix of the silhouette is calculated by using FEA. The stiffness matrices are represented as feature vectors. The RBF-SVM classifier is used for the classification of actions.

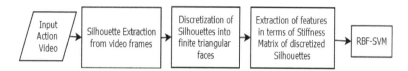

Figure 2.1 Workflow diagram of the proposed method

2.2.1 Silhouette Extraction

In the proposed method, features are acquired to distinguish different human actions which make the result more authentic. These features describe the deformation that occurs in the silhouette in terms of shape and motion information while performing an action. As silhouette moves, the finite elements also get deformed. The stiffness matrix of the silhouette narrates these features. The first step of the proposed method is silhouette extraction which is also a very challenging problem because it requires background subtraction. Background cluttering, illumination change, noises, etc. are some challenges for background subtraction. The GMM [44] is robust to problems discussed above and it also has the capability to deal with the critical issue like a shadow. We used GMM for background subtraction. Then the silhouette is extracted and normalized so that all the silhouettes become equal in size [37]. Figure 2.2 shows the extracted silhouette from the video.

Figure 2.2 Walk and the extracted silhouette

2.2.1 Discretization and Shape Function Representation

The preliminary step for FEA is discretization i. e. modeling the silhouette structure into numbers of small elements as shown in Figure 2.3. The number of elements in which geometry is divided is variable and can be determined by software like MATLAB, COMSOL, etc. which demand physics of the geometry. We selected the simple triangular element as the finite element. The reason behind using the triangular shape is that it is the simplest structure for numerical representation. We used MATLAB having an FEA toolbox in the proposed method.

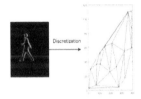

Figure 2.3 Segmented discretized silhouette

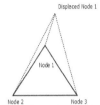

Figure 2.5 Deformed triangle as node 1 is displaced

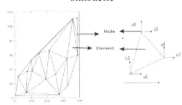

Figure 2.4 Displacement of a mesh (triangle) of the human body and its mesh element

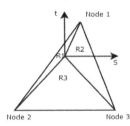

Figure 2.6 Division of the triangle into 3 parts

We referred Laptev et al. [25] to find out the prominent points at the boundary of the silhouette. These prominent points are given as the nodes to the FEA toolbox. The silhouette is discretized into the finite triangular elements using these nodes as vertices of the triangle. The discretization of the silhouette is done in such a way that they do not overlap. In Figure 2.3 silhouette is discretized into finite triangular elements. The discretized structural representation of the silhouette is shown on the right side of Figure 2.3 where X-axis and Y-axis are the spatial coordinates of the vertices of the triangle. The complete domain is divided into simpler parts; it provides precise and detailed representation for analysis. Each triangular element has three nodes. Every node has a displacement in the X direction and Y direction as shown in Figure 2.4. Displacement vectors of the triangle can be represented in equation 2.1.

$$U = \{u_1, u_2, u_3, u_4, u_5, u_6\}^T \qquad (2.1)$$

where, U is a displacement vector for each triangular element and u_1 and u_2 are the displacement of node 1 in X and Y direction respectively, u_3 and u_4 are the displacements of node 2 in X and Y direction and similarly u_5 and u_6 are the displacements of node 3.

When silhouette moves from one frame to another frame, nodes of the triangles are also displaced. To build the face correspondence between the frames we calculated the Euclidean distance among the nodes of the previous frame and next frame. The minimum distance will correspond to the same points. The displacement of the nodes of the triangle is found out from one frame to another frame. Shape function of the triangle is used to represent the nodal displacement. Figure 2.5 shows the displacement of node 1 as a dotted line from its previous point to the displaced point, which results in the deformation of the triangle assuming that the other two nodes are fixed. Similarly, node 2 and node 3 are considered. An interior-point is taken inside the triangle to divide it into 3 regions as shown in Figure 2.6 Let the total area of the triangle is R and R_1, R_2 and R_3 are the areas of three regions. This is represented by Equation 2.2.

$$R = R_1 + R_2 + R_3 \tag{2.2}$$

From Equation 2.2 the shape functions of all three regions are evaluated using Equation 2.3.

$$J_1 = \frac{R_1}{R}, \ J_2 = \frac{R_2}{R} \text{ and } J_3 = \frac{R_3}{R} \tag{2.3}$$

where J_1, J_2, and J_3 are the shape functions of regions R_1, R_2, and R_3. We assume the displacement of the interior point of the triangle in Figure 2.3.d is s and t in X and Y directions respectively. Displacements s and t can be represented with the help of the shape functions by Equation 2.4 and Equation 2.5.

$$s = J_1 u_1 + J_2 u_3 + J_3 u_5 \tag{2.4}$$

$$t = J_1 u_2 + J_2 u_4 + J_3 u_6 \tag{2.5}$$

In Equation 2.4 and Equation 2.5, u_1, u_3, and u_5 are the displacement in the X direction and u_2, u_4 and u_6 are the displacements in the Y direction of the vertices of the triangle. The shape functions are determined by the areas of the different regions of the triangle, thus, the regions of the triangle are dependent on each other and can be represented by Equation 2.6.

$$J_1 + J_2 + J_3 = 1 \tag{2.6}$$

From Equation 2.6, we can say that if we know two shape functions, then the third function can be easily calculated from Equation 2.7.

$$J_3 = 1 - J_1 - J_2 \tag{2.7}$$

Putting these values to Equation 2.4 and Equation 2.5 we get the displacement of the interior point of the triangle in terms of s and t in X and Y directions respectively as shown in Equation 2.8 and Equation 2.9.

$$s = (u_1 - u_5) J_1 + (u_3 - u_5) J_2 + u_5 \qquad (2.8)$$

$$t = (u_2 - u_6) J_1 + (u_4 - u_6) J_2 + u_6 \qquad (2.9)$$

2.2.3 Representation of Feature Vector

The displacement of the interior point represents the displacement of the triangle. As discussed above the interior point (x, y) has displacements s in X direction and t in the Y direction. Due to these displacements, a deformation is produced in the triangular element. This deformation is nothing but the strain developed in the triangular element in X, Y, and shear direction. These are given in Equation 2.10, Equation 2.11 and Equation 2.12 respectively.

$$\text{Strain in X-direction: } \phi_x = \frac{\partial s}{\partial x} \qquad (2.10)$$

$$\text{Strain in Y-direction: } \phi_y = \frac{\partial t}{\partial y} \qquad (2.11)$$

$$\text{Shear strain: } \phi_{xy} = \frac{\partial s}{\partial y} + \frac{\partial t}{\partial x} \qquad (2.12)$$

These strains are written in the form of matrices in Equation 2.13.

$$\phi = \begin{bmatrix} \frac{\partial s}{\partial x} \\ \frac{\partial t}{\partial y} \\ \frac{\partial s}{\partial y} + \frac{\partial t}{\partial x} \end{bmatrix} \qquad (2.13)$$

Once we get the strain in the triangular element of the discretized silhouette, we found out the stiffness matrix by Equation 2.14 using FEA [123].

$$k_t = C^T D C \, t_e \, R_e \qquad (2.14)$$

where k_t is the stiffness matrix for a triangle element, C is a displacement matrix subject to strains in X direction, Y direction, and shear strain, t_e is thickness of the body which is constant in case of silhouette, R_e is the area of the triangle and D is a constant matrix represented by Equation 2.15.

$$D = \frac{\xi}{1-\tau} \begin{bmatrix} 1 & \tau & 0 \\ \tau & 1 & 0 \\ 0 & 0 & \frac{1-\tau}{2} \end{bmatrix} \quad (2.15)$$

where ξ is Young's modulus and τ is Poisson's ratio and both are constants. We tuned these parameters and we discussed their values in the experimental result section. Further, we converted the stiffness matrix of the triangle k_t into a one-dimensional feature vector [37] by scanning the matrix from the top left to bottom right element by element. The stiffness matrix of the triangle having m rows and m columns will be converted into the one-dimensional feature vector having total *m* x *m* elements. Similarly, we calculated the feature vectors of all possible triangles of the silhouette. The complete stiffness matrix of the silhouette K_s is created by combining all feature vectors of triangles where rows of the matrix represent the triangles associated with the silhouette. To solve the issue that which feature vector of the triangle will be the first row of the Stiffness Matrix of the silhouette we adopted the following strategy:

We scanned the discretized silhouette from top left to bottom right (interior point of the triangle). The first triangle whose interior point is found first in scanning will represent the first row of the matrix K_s and the triangle whose interior point is found last will represent the last row of the K_s matrix. If a silhouette of a frame is discretized into n number of triangle face then the stiffness matrix of silhouette K_s will have n number of rows and *m* x *m* numbers of the column. Further, the complete stiffness matrix of a silhouette is converted into the feature vector with a similar procedure as discussed above. This feature vector represents the frame of the action video.

2.2.4 Dimension Reduction and Classification

A frame of an action video at time t is represented by the feature vector extracted from the proposed method. The length of the feature vector of a frame is $C = row \times column$ of the stiffness matrix of the silhouette. Suppose an action sequence consists of S frames, then that action sequence has S feature vectors. This results in a very high dimensional feature space. To reduce the dimensional feature space, we applied Principal component analysis (PCA). Further, these reduced features are given to RBF-SVM classifier [103 and 104] to recognize the actions.

2.3 Algorithm

The proposed methodology can be summarized in the form of the algorithm as follows

Step 1: Extraction of silhouettes from input video frames.

Step 2: Extraction of prominent points on the boundary of the silhouettes.

Step 3: Prominent points are given as nodes to the FEA toolbox, MATLAB.

Step 4: Silhouettes are discretized into finite triangular elements where nodes act as vertices of the triangle.

Step 5: Each triangular element is represented by three nodes displacement vector (U) of the triangle.

Step 6: Displacement matrix (C) of each triangle is calculated.

Step 7: Stiffness matrix (kt) is calculated for each triangle with the help of displacement matrix C.

Step 8: Complete the Stiffness matrix of the silhouettes is created by combining stiffness matrices (kt) of all possible triangles of the silhouette.

Step 9: Stiffness matrix of the silhouette is represented as one-dimensional feature vectors

Step 10: Feature vector is calculated for all frames of the action video for all actions.

Step 11: The RBF-SVM Classifier is used for recognition.

2.4 Experimental Setup

We have developed our proposed method on MATLAB R2015a. The proposed algorithm has been tested on a system having hardware configuration processor Intel(R) Core (TM) i5-6200U CPU @2.30GHz 2.40 GHz with 8 Gb RAM and 64-bit operating system. To evaluate the performance of the proposed methodology, accuracy is used as the performance parameter in a leave-one-out cross-validation strategy. It can be represented by using a true positive rate (TPR) and a false-positive rate (FPR) represented by Equation 2.16 and Equation 2.17.

$$TPR = \frac{(True\ Positive)}{(True\ Positive + False\ Negative)} \qquad (2.16)$$

$$FPR = \frac{(False\ Positive)}{(False\ Positive + True\ Negative)} \qquad (2.17)$$

where TPR represents positive cases that are correctly classified and FRP represents negative cases that are incorrectly classified as positive. Accuracy is calculated by Equation 2.18.

$$Accuracy = \frac{True\ Positive + True\ Negative}{True\ Positive + True\ Negative + False\ Positive + False\ Negative} \qquad (2.18)$$

2.4.1 Datasets Used

We have chosen four challenging datasets for action recognition namely Weizmann dataset [19], KTH [105], Ballet [45] and IXMAS [46] to evaluate and compare our proposed method.

Weizmann Dataset: In the Weizmann action dataset, there are ninety videos. The frame rate is 25 frames per second (fps) and resolution is *144×180* pixels. It consists of nine different persons who performed a total of ten actions such as running, jumping, waving, bending, etc. The sample frame is shown in Figure 2.7

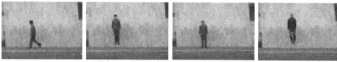

Figure 2.7 Weizmann dataset

KTH Dataset: The KTH dataset comprises six essential exercises, in particular; applauding, waving, boxing, walking, jogging and running. Activities in KTH have been recorded in four different lightings, indoor and outdoor situations and have 100 recordings. But the foundation for all recordings has been kept the same with a static camera with 25 fps and resolution of *160* x *120* pixels. The states of the recordings in the KTH informational index suffer from camera development and lighting impacts. The sample frame of this dataset is shown in Figure 2.8

Figure 2.8 KTH dataset

Ballet Dataset: Ballet is an expressive dance dataset, which comprises of profoundly complex artful dance poses of various on-screen characters. The specimen casings of the dataset are shown in Figure 2.9. The dataset is acquired from a ballet dance DVD. The foundation in the dataset is basic. Every video grouping comprises of just a single performing artist. The dataset comprises of 44 videos. There are eight different unique activities performed in these videos.

Figure 2.9   Ballet dataset

IXMAS Dataset: IXMAS is a very challenging dataset where 10 distinctive persons are performing every activity three times. These videos have been recorded from different view perspectives where seven different cameras used for recording. These activities incorporate scratching head, looking at the watch, strolling, taking a seat, etc. This dataset offers different challenges by introducing huge appearance change, intra-class varieties, and self-impediments, etc. XMAS results have been evaluated for five different camera views. The sample from the IXMAS dataset is shown in Figure 2.10.

Figure 2.10   IXMAS dataset (5 cameras)

2.4.2 Parameter Settings

For the parameter settings, we tuned the important parameters on the KTH dataset and similar settings are applied to other datasets. These important parameters are the number of nodes, number of finite elements and feature dimension through PCA. The number of nodes is the prominent points extracted on the silhouette boundary. We have experimented on 5, 10, 15, 17, 20 and 25 prominent points which were considered as nodes. It is clear from Figure 2.11.a that when a few prominent points are greater than 15, we achieve a good result. In the proposed method we have taken 17 numbers of nodes because accuracy is varying only 1-2% as we take several points greater than 17. A next parameter is several finite elements. The number of finite elements has experimented as 10, 15, 20, 22 and 25. Figure 2.11.b shows that the number of elements greater than 20 is giving better accuracy. The more discretized element we have, the more will be the accuracy of the representation of the body structure. But the trade off is that more discretized elements will increase the complexity in terms of time. Thus, we have taken 22 numbers of triangular faces in the proposed method. These 22 numbers of finite elements are taken in such a manner that these triangles do not overlap. This makes the structure simple and attains better accuracy.

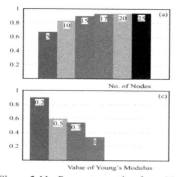
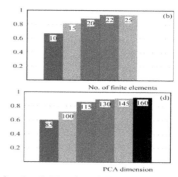

Figure 2.11 Parameter setting for a. Number of nodes; b. Number of finite elements; c. Young's Modulus; d. Feature dimension through PCA

As far as Young's modulus mentioned in Equation 2.15 is concerned, we have experimented on its normalized values 0.2, 0.5, 0.7 and 1.0 as shown in Figure 2.11.c. We got the highest accuracy when the value of Young's Modulus was 0. 2. The possible reason behind it could be the value of Youngs modulus is higher for the rigid body and lower for the flexible body. As the human body is very flexible while performing an action, the lower value 0.2 gives a better result. The Poisson's ratio mentioned in Equation 2.15 is used for the material property and it is a constant value that lies between 0-0.5. In the proposed method we got the optimized result when the value of was 0. 5. The thickness of the silhouette discussed in Equation 2.14 remains constant for all frames in a video and in the proposed methodology; we have taken the value of thickness as 1. The last parameter is the feature dimension through PCA. The result of PCA for different dimensions is represented in Figure 2.11.d. We have experimented on different values of dimension such as 85, 100, 115, 130, 145 and 160. Here dimension 130 is showing better results in terms of accuracy and complexity.

2.4.3 Results Discussion

We applied the leave-one-out strategy for cross-validation. Table 2.1, Table 2.2, Table 2.3 and Table 2.4 shows the confusion matrices resulted from applying the proposed method on the datasets Weizmann, KTH, Ballet, and IXMAS respectively. These confusion matrices show that most of the actions are 100% classified except some similar types of actions. Thus, we got an accuracy of 97.8%, 96.4%, 95.2% and 90.3% for the Weizmann, KTH, Ballet, and IXMAS respectively. We compared the proposed method with the other Silhouette analysis based human action recognition methods such as

Human Body Pose Model (HBPM) [35, 36 and 57] and Human Body Pose Temporal Model (HBPTM) [37, 47] for these datasets. Chaaraoui et al. [57] extracted the features of the silhouette as contour points and action is learned from the multi-views of cameras. The multi-view learning makes the method capable of differentiating different persons performing the same action. Wu et al. [37] proposed the Human Body Pose Temporal Model where a 2-D silhouette mask is converted into a 1-D feature vector. They represented the action as the correlogram of poses extracted from the silhouette. H. Han et al. [60] represented the human body shapes with sparse geometrical features using the Bandlets transformation. They used the AdaBoost to select the features.

Table 2.1. Confusion matrix for Weizmann Dataset (R-Running, W-Walking, J-Jumping, JJ-Jumping Jack, S-Skipping, JP-Jumping at a place, SJ-Side Jumping, B-Bending, W-Waving with one hand, WB-Waving with both hands)

	R	W	J	JJ	S	JP	SJ	B	W	WB
R	0.95	0.05	0	0	0	0	0	0	0	0
W	0	1	0	0	0	0	0	0	0	0
J	0	0	1	0	0	0	0	0	0	0
JJ	0	0	0	1	0	0	0	0	0	0
S	0	0	0	0	0.96	0.04	0	0	0	0
JP	0	0	0	0	0.02	0.98	0	0	0	0
SJ	0	0	0	0	0	0	1	0	0	0
B	0	0	0	0	0	0	0	1	0	0
W	0	0	0	0	0	0	0	0	1	0
WB	0	0	0	0	0	0	0	0	0	1

Table 2.2. Confusion matrix for KTH Dataset (A- Applauding, W- Waving, B- Boxing, WK- Walking, J- Jogging, R- Running)

	A_1	A_2	A_3	A_4	A_5	A_6
A	1	0	0	0	0	0
W	0	1	0	0	0	0
B	0	0.02	0.98	0	0	0
WK	0	0	0	1	0	0
J	0	0	0	0	0.96	0.04
R	0	0	0	0	0.02	0.98

Table 2.3. Confusion matrix for Ballet LR- Left to right-Hand Opening, RL- Right to left-Hand Opening, J-Jumping, H- Hopping, S- Swinging leg, ST- Standing, T- Turning)

	A_1	A_2	A_3	A_4	A_5	A_6	A_7
LR	1	0	0	0	0	0	0
RL	0	1	0	0	0	0	0
J	0	0	0.97	0.03	0	0	0
H	0	0	0.10	0.91	0	0	0
S	0	0	0	0	1	0	0
ST	0	0	0	0	0	1	0
T	0	0	0	0	0	0	1

Table 2.4. Confusion matrix for IXMAS Dataset (W- Walking, WA- Waving, P- Punching, K- Kicking, T- Throwing, P- Pointing, PU- Picking Up, G- Getting Up, S- Sitting Down, TA-Turning Around, F-Folding arms, C-Checking Watch, SH-Scratching Head)

	W	WA	P	K	T	P	PU	G	S	TA	F	C	SH
W	1	0	0	0	0	0	0	0	0	0	0	0	0
WA	0	0.92	0.08	0	0	0	0	0	0	0	0	0	0
P	0	0	0.97	0	0.03	0	0	0	0	0	0	0	0
K	0	0	0	1	0	0	0	0	0	0	0	0	0
T	0	0	0.03	0	0.94	0.03	0	0	0	0	0	0	0
P	0	0	0	0	0.03	0.97	0	0	0	0	0	0	0
PU	0	0	0	0	0	0	0.94	0.06	0	0	0	0	0
G	0	0	0	0	0	0	0.8	0.92	0	0	0	0	0
S	0	0	0	0	0	0	0	0	1	0	0	0	0
TA	0	0	0	0	0	0	0	0	0	1	0	0	0
F	0	0	0	0	0	0	0	0	0	0	0.97	0.03	0
C	0	0	0	0	0	0	0	0	0	0	0.04	0.94	0
SH	0	0	0	0	0	0	0	0	0	0	0	0	1

As the proposed method gives the precise change in the human body shape due to the change in the small elements of the silhouette, it makes it better as compared to other methods. Figure 2.12 shows that the proposed method shows a better result as compared to other silhouette analysis based methods using Weizmann, KTH, Ballet, and IXMAS datasets.

We also compared other state-of-the-art methodologies with the proposed method on all four datasets in Table 2.5-Table 2.8. Different testing strategies are used in these methodologies. We mentioned these testing strategies in the tables along with the classifier that they have used. Table 2.5 shows a comparison of the proposed method with other methodologies on the Weizmann dataset. Goudelis et al. [58] proposed a new feature extraction technique based on the Trace Transform. They represented the spatiotemporal feature in terms of the trace transform from the binarized silhouette. They used the SVM classifier and leave-one-person-out cross-validation testing strategy. They achieved 94.6% accuracy.

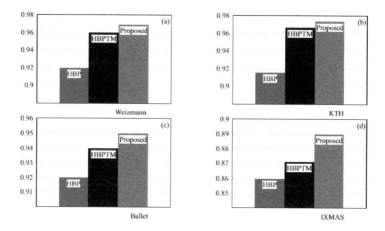

Figure 2.12 Comparison of the proposed method with similar methods for four datasets ...Weizmann, KTH, Ballet, and IXMAS

The methods [37] and [63] achieved a higher accuracy of 96.3% and 97.3% respectively. Liu et al. [63] modeled human action with the neural mechanism. They used new feature vectors using two cortical areas one is the primary cortex and the second is the middle temporal cortex for motion. Later, they used the SVM classifier to recognize the action. The proposed method achieved an accuracy of 97.8%. Since we have discretized the silhouette into a smaller triangle, similar actions such as walking and running are recognizable better than the other methods. Unlike the Weizmann dataset, the KTH dataset offers more challenging environments. It has different setups having different lighting conditions for different actions. Moreover, the shadow is also a very big challenge in this dataset. So to deal with these problems, GMM is a better strategy for background subtraction and silhouette extraction. Rahman et al. [47] used the negative space-based feature of human pose and motion features to model the actions. To classify the actions they used Nearest Neighbor Classifier. The leave-one-out strategy is used for cross-validation. They showed an accuracy of 95.1% as shown in Table 2.6. In another method, Shi et al. [64] proposed new motion descriptor sequential deep trajectory descriptors for long term motion video. The CNN-RNN network is used to learn the motion. They achieved a comparable accuracy of 95.8% as compared to [47]. We have used a leave-one-out strategy and the proposed method achieved 96.4% accuracy which is better than other methods.

Abbreviations used in Table 2.5-Table 2.8 are SVM: Support Vector Machine, KNN: K-Nearest Neighbor, LOSO: Leave-One-Sequence-Out, LOPO: Leave-One-Person-Out, LOO: Leave-One-Out, NNC: Nearest Neighbor Classifier, CNN: Convolutional Neural Network, HMM: Hidden Markov Model, RNN: Recurrent Neural Network, SVM-NN: Support Vector Machine-Neural Network, LOOCV: Leave-One-Out Cross-Validation, S-CTM: , RVM: Relevance Vector Machine.

Table 2.5 Comparison of the proposed method with similar methods on Weizmann dataset

Method	Year	Classifier and Test	Accuracy
[60]	2015	SVM	81.5
[57]	2013	KNN, LOSO	91.7
[37]	2013	SVM, LOSO	96.3
[58]	2013	SVM, LOPO	94.6
[59]	2014	KNN, LOO	91.4
[61]	2017	NNC	95.3
[62]	2016	CNN-HMM	90.1
[63]	2017	SVM	97.3
Proposed Method		SVM, LOO	97.8

Table 2.6 Comparison of the proposed method with similar methods on KTH dataset

Method	Year	Classifier and Test	Accuracy
[60]	2015	Adaboost, SVM, LOO	94.2
[61]	2017	SVM	93.8
[58]	2013	SVM, LOPO	92.7
[61]	2017	NNC, LOO	93.6
[62]	2016	CNN-HMM	94.4
[63]	2017	SVM	91.3
[47]	2014	KNN, LOO	95.1
[48]	2015	KNN	90.8
[64]	2017	CNN-RNN	95.8
Proposed Method		SVM, LOO	96.4

Table 2.7 Comparison of the proposed method with similar methods on Ballet dataset

Method	Year	Classifier and Test	Accuracy
[35]	2017	SVM-NN, LOOCV	94.2
[36]	2015	SVM-NN, LOOCV	93.8
[49]	2009	S-CTM, LOO	89.8
[50]	2014	RVM, LOO	90.4
[51]	2014	SVM, LOO	90.3
Proposed Method		SVM, LOOCV	95.2

Table 2.8 Comparison of the proposed method with similar methods on IXMAS dataset

Method	Year	C1	C2	C3	C4	C5	Overall Accuracy
[52]	2011	89.1	83.4	89.3	87.2	89.2	87.8
[53]	2010	84.2	85.2	84.1	81.5	82.6	82.7
[54]	2013	86.5	83.8	86.1	84.5	87.4	87.2
[55]	2016	91.3	85.7	89.3	90.2	86.5	87.5
Proposed Method		90.8	90.6	92.4	91.2	90.6	90.2

Table 2.7 shows a comparison of the proposed method with the other state-of-the-methods for the Ballet dataset. Vishwakarma et al. [36] used silhouette based analysis

and extracted the feature vectors based on human poses. They have used SVM, LDA and Neural Network-based hybrid classifiers to recognize the action. They achieved an accuracy of 93.8%. Vishwakarma et al. [35] used a new silhouette analysis where they first found out the Average Energy Image of a silhouette. The spatial distribution of gradient is applied on Average Energy Image to make it a global feature and the temporal feature is found out by Radon transform of the silhouettes. These features are given to the hybrid classifier and they achieve a higher accuracy of 94.2%. In both methods [35] and [36] they used leave-one-out cross-validation. The proposed method shows better accuracy of 95.3%. As discussed above discretized silhouette into small triangles helps to recognize the actions in an expressive dataset like Ballet dance. In this dataset, the closely the performer's expressions are observed the better results could be achieved. IXMAS dataset has five different camera views. Methods [50, 54 and 55] show almost similar accuracies which are around 87%. Wang et al. [55] used the Bag-of-visual-word method based on local features. Then they used the cross-view approach to deal with the problem of view change due to different cameras. The proposed method has achieved higher Accuracy for all the views. We got an average of 90.2% accuracy. To deal with the problem of variation in viewpoint we used view-invariant interest point on the boundary of the silhouette which acted as the vertices of the triangles during the discretization step. Table 2.8 shows a comparison of the proposed method the other state-of-the-methods for IXMAS dataset

2.5 Run Time Analysis

For run-time analysis, we have used NVIDIA GPU and MATLAB 2015a with a Parallel computing toolbox. Time taken for several modules in the proposed methodology is calculated. We have analyzed the run-time of the proposed method on all the datasets discussed above. The average run-time of the proposed dataset and is given below step-wise:

1. Extraction of Silhouette from Action Video (Sec): 0.41
2. Silhouette discretization into Triangular faces (Sec): 0.54
3. Calculation of stiffness matrix of the Silhouette (Sec): 1.45
4. Classification (Sec): 0.52

Total time for Action Recognition (Sec): 2.92

Thus, the run-time of the proposed method is fairly good.

2.6 Chapter Conclusion

This is a new method to recognize human action through Finite Element Analysis (FEA). A new feature descriptor where the feature vectors of the video frames are expressed in terms of the stiffness matrix of the silhouette extracted from the frames of the video is applied. This offers uniqueness to this method, as it can extract both shapes as well as motion information. The feature vectors extracted from the proposed method are given to the RBF-SVM classifier. Validation of the proposed method has been performed in different challenging environments. The limitation of the methodology is that it requires accurate silhouette extraction. The proposed method shows its superiority as compared to other existing methods of applying them on challenging standard datasets such as Weizmann, KTH, Ballet, and IXMAS.

3. MODIFIED BAG-OF-VISUAL-WORDS BASED DESCRIPTOR

3.1 Introduction

The actions are described by the global and local feature descriptors. The global feature descriptor gives shape as well as motion information of the person who is performing an action. These are not invariant to the appearance change, shape changes, etc. whereas local descriptor such as spatio-temporal points show invariance in appearance change, background change, etc. Local features based on Bag-of-visual–words have been very popular among the researchers. The disadvantage of these methods is that they do not retain the structural information of the clusters formed in Bag-of Visual words based methodology. The proposed method overcomes the limitation of the Bag-of-visual-word methodology. The proposed methodology contributed in the following manner:

I. The contextual distance among the points of a cluster is calculated on the basis of the difference between the contributions of points to maintain the geometrical structure of the cluster, unlike the contextual distance calculated on basis of Euclidean measure among the points in a cluster.

II. The contextual distance among any two points of the contextual set of a cluster, calculated as mentioned above may show asymmetry. So to deal with this asymmetry issue, we used the concept of the directed graph. Thus, a cluster is represented using contextual distance and directed graph together in a robust way.

3.2 Proposed Methodology

Local features such as spatiotemporal points are most popularly used to represent an action. Spatiotemporal interest points such as Harris 3D detectors and SIFT detectors [14] are used to detect the interest points. We used the Harris 3D detector to extract the spatio temporal interest points. These interest points are described locally by the descriptors such as HOG [132'] and HOF [51]. The HOG and HOF are used in the proposed method to describe the interest points detected by Harris 3D detector. These local descriptors represent the feature vectors of each frame of the action video. A bag-of-visual-word of spatiotemporal interest points is formed to represent an action efficiently. The codebook is created by performing vector quantization through clustering. The K-means clustering [135] is popularly used. In K-means clustering clusters are placed near to most frequently occurring feature vectors. These most frequently occurring feature vectors are used as the discriminative features of action. But for certain actions, these most frequently occurring features are not capable enough to show the discriminative nature of an action. Further quantization error may also occur due to the assignment of a single cluster to feature vector. To avoid this we used the soft assigning technique [37] where Gaussian kernel is used in which feature vectors and cluster are represented by the normal distribution. Equation 3.1 defines the visual word $V_{i,t}$

$$V_{i,t} = \exp\left(-\frac{\|f_t - C_i\|^2}{2\theta^2}\right) \quad (3.1)$$

where, f_t is the feature vector and C_i is the ith cluster where i=1, 2, 3... N and θ is the smoothing parameter. We used the cross-validation to find out the size of the Gaussian kernel. In the proposed method we used this soft assigning technique and we get the codebook $\{C_1\ C_2\ C_3\ ...\ C_N\}$ where C_i is the ith cluster and the total number of the cluster are N. Figure 3.1 shows the proposed methodology for codebook creation. Let the spatiotemporal feature points be represented in the form of tuple X = (L, r) where L is local descriptor (appearance) of the interest points and r = (x, y, t) is the spatial location of the interest points. The horizontal, vertical and temporal coordinates of interest points are represented by x, y, and t respectively. In the traditional bag-of-visual-word based methodology, the geometrical structure of the cluster cannot be maintained because of the introduction of noise. This results in the reduction of the recognition rate. Thus, to maintain the geometrical structure of the clusters, we propose a new methodology where we calculate the contextual distance among the points of the cluster on the basis of the difference between the contributions of points in a cluster to maintain its geometrical structure. The Methodology can be understood by the block diagram in Figure 3.1. The spatiotemporal points are detected from the input action video by using Harris 3D corner detector. These points are described by the descriptors HOG and HOF. Using soft assigning technique we created the codebook having N no. of clusters. The traditional bag-of-visual-word technique is modified through the proposed methodology. Using the geometrical descriptor, the contextual distance among the points of the cluster is calculated. The directed graph of the cluster points is then created with the help of contextual distance between the points. These directed graphs are described by Laplacian.

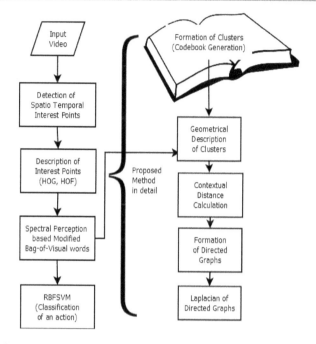

Figure 3.1 Detail process flow diagram of the proposed methodology

3.2.1 Geometrical Description of a Cluster

As discussed above, among the N clusters formed, let us select the ith cluster C_i. The cluster C_i is represented by the set I_c. The set I_c is represented by Equation 3.2 where i_c is a mean point of the cluster C_i having m nearest neighbouring points.

$$I_c = (i_{c0}, i_{c1}, i_{c2}, ..., i_{cm}) \qquad (3.2)$$

We used centroid as the geometrical structure descriptor of set I_c denoted by Equation 3.3 as shown below,

$$M(I_c) = \frac{1}{m+1} \sum_{j=0}^{m} i_{ck} \qquad (3.3)$$

To calculate the contribution of each point in the set I_c, we removed one point at a time and recalculated the centroid using Equation 3.3. Thus we get the contribution of all points in the set I_c that forms the geometrical structure of the cluster. The contribution of point i_{ck} can be found out by Equation 3.4.

$$\delta(i_{ck}) = \left| M(I_c) - M\left(\frac{I_c}{i_{ck}}\right) \right| \tag{3.4}$$

where $\delta(i_{ck})$ is the contribution of the point i_{ck} in the cluster set I_c. $M(I_c)$ is the complete descriptor of I_c and $M(I_c/i_{ck})$ is the description of set I_c after removing the point i_{ck} from the set. In a similar manner contribution of all the points in a cluster's set is found out. If the value of $\delta(i_{ck})$ is very small then the contribution of the point i_{ck} is not significant in maintaining the geometrical structure of the cluster and if the value of $\delta(i_{ck})$ is high then we can conclude that the point i_{ck} has a significant effect in retaining the geometrical structure of the set I_c.

3.2.2 Contextual Distance among the points of a Cluster

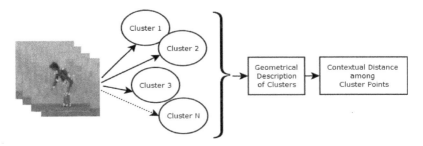

Figure 3.2 Geometrical description of a cluster and contextual distance among the cluster points

Figure 3.2 describes the geometrical description of the cluster and the contextual distance among the points of the clusters. Once we get the contribution of points of a cluster to maintain its geometrical structure, we found out the contextual distance among these points by taking the difference between their contributions. This can be calculated from Equation 3.5.

$$\Psi(i_{c0}, i_{c1}) = \{\delta(i_{c0}) - \delta(i_{c1})\} \tag{3.5}$$

where $\Psi(i_{c0}, i_{c1})$ is the contextual distance between interest points i_{c0} to i_{c1} and $\delta(i_{c0})$ and $\delta(i_{c1})$ are the contribution of the points i_{c0} and i_{c1}. After calculating the contextual distances among the points of the cluster set I_c we can represent the set I_c in the form of their contextual distances. This representation of the cluster set creates asymmetry among the points of the set I_c because the contextual distance between i_{c0} to i_{c1} may not be equal to i_{c1} to i_{c0}. To deal with this naturally occurred asymmetry in the cluster set I_c, we used directed graph which made the cluster more informative as it introduced direction also.

3.2.3 Formation of the Directed Graph of a Cluster on the Basis of Contextual Distance

The cluster set I_c can be represented by the directed graph where the points of set are treated as the nodes/vertices of the graph as shown in Figure 3.3.

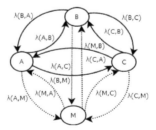

Figure 3.3 A cluster set I_c of m points where A=i_{c0}, B=i_{c1} and λ (A, B) = (i_{c0}, i_{c1})

Let the edge from the points i_{c0} to i_{c1} is defined by the weight $\lambda(i_{c0}, i_{c1})$ as given by Equation 3.6.

$$\lambda(i_{c0}, i_{c1}) = e^{-\left[\frac{\Psi(i_{c0}, i_{c1})}{\mu}\right]^2} \tag{3.6}$$

where $\lambda(i_{c0}, i_{c1})$ is the weight of the edge from point i_{c0} to i_{c1}, $\Psi(i_{c0}, i_{c1})$ is the contextual distance from i_{c0} to i_{c1}, μ is a free parameter and i_{c0} and i_{c1} are the neighbourhood points of the cluster set I_c. Thus, in the cluster set I_c a point i_c is connected to its m neighbourhood points through m edges as shown in figure 3.3.

The directed graph of cluster set can be represented through weighted edge matrix W. The weighted edge matrix W carries the geometrical structural information of the cluster set I_c. Similarly, we applied a directed graph on other contextual clusters sets also. The weighted edge matrix is created for a single cluster as shown below:

$$\begin{bmatrix} & i_{c0} & i_{c1} & i_{c2} & \cdots & i_{cm} \\ i_{c0} & \lambda(i_{c0}, i_{c0}) & \lambda(i_{c0}, i_{c1}) & \lambda(i_{c0}, i_{c2}) & \cdots & \lambda(i_{c0}, i_{cm}) \\ i_{c1} & \lambda(i_{c1}, i_{c0}) & \lambda(i_{c1}, i_{c1}) & \lambda(i_{c1}, i_{c2}) & \cdots & \lambda(i_{c1}, i_{cm}) \\ i_{c2} & \lambda(i_{c2}, i_{c0}) & \lambda(i_{c2}, i_{c1}) & \lambda(i_{c2}, i_{c2}) & \cdots & \lambda(i_{c2}, i_{cm}) \\ \vdots & \cdots & \cdots & \cdots & \ddots & \vdots \\ i_{cm} & \lambda(i_{cm}, i_{c0}) & \lambda(i_{cm}, i_{c1}) & \lambda(i_{cm}, i_{c2}) & \cdots & \lambda(i_{cm}, i_{cm}) \end{bmatrix}$$

Further to describe the weighted directed graph we used Laplacian of the directed graph

3.2.4 Laplacian of the Directed Graph of a Cluster

Suppose G (i_{cj}, λ), where j = 0...m, is the directed graph of the set I_c and the directed edge (i_c, i_d) is the edge directed from vertex i_c to vertex i_d. A walk from one vertex to other is represented by the transition probability matrix O. If $O(i_c, i_d)$ is the transition probability of moving from point i_c to i_d, it is obvious that $O(i_c, i_d) > 0$ only if there is an edge between i_c to i_d. Further for the weighted directed graph, the transition probability matrix is given by Equation 3.7.

$$O(i_c, i_d) = \frac{\lambda_{(i_c, i_d)}}{\sum_m \lambda_{(i_c, i_m)}} \qquad (3.7)$$

wherein Equation $\lambda_{(i_c, i_d)}$ is the weight of the edge from point i_c to i_d and denominator in the right-hand side of the equation represents the total weighted sum of all the edges from point i_c to all its neighboring points.

According to Equation 3.7 transition probability matrix of the directed graph has a unique left Eigenvectors ε with all positive values and the Laplacian of the transition probability matrix is defined by the Equation 3.8

$$\Phi = I - \frac{\varepsilon^{\frac{1}{2}} O \varepsilon^{-\frac{1}{2}} + \varepsilon^{-\frac{1}{2}} O \varepsilon^{\frac{1}{2}}}{2} \qquad (3.8)$$

where O is the transition probability matrix and ε is the Eigenvector. Equation 3.8 gives the Laplacian of the transition probability matrix for a cluster C_i. This Laplacian of the transition probability matrix is converted to a vector having dimension by scanning the matrix from the top left element to bottom right element. Thus, Laplacian of the transition probability matrix for N clusters has N feature vectors. These feature vectors are fed to RBF-SVM classifier. Figure 3.4 shows formation of N clusters and their Laplacian descriptors.

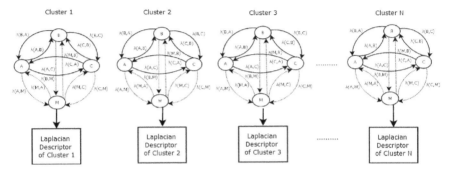

Figure 3.4 Formation of N clusters and their Laplacian descriptors

3.3 Parameter Setting

There are two parameters that can affect the proposed methodology. The First parameter is the size of codebook using the soft assigning strategy and second is the number of nearest neighbors of a cluster. In most of the bag-of-visual-word based methods, the size of the codebook is around 1000. But in the proposed method we have done cross validation optimization to find out the size of codebook. Figure 3.5 shows the codebook size 30, 40, 50 and 60. We opted the codebook size 35 because if the size of the cluster is greater than 30 then the accuracy varies only 2-3%. Another parameter is the number of nearest neighbour in the cluster. We have taken different values such as 100, 120, 140, 160, 180, 200 and 220. We have taken the size of nearest neighbour 160 because other than this value the accuracy varies only 1-2% as shown in Figure 3.6. Similar setting is used for KTH and Ballet datasets.

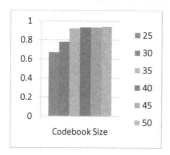 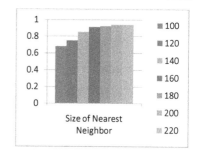

Figure 3.5 Selection of codebook size   Figure 3.6 Selection of size of nearest neighbor

3.4 Results

3.4.1 Analysis and Discussion on KTH

Figure 3.7 shows the confusion matrix where we get the appreciable result for differentiating different actions. Similar types of actions like jogging and running are also very well classified. Only 5% jogging action is wrongly classified as running and running is only 3% confused with the jogging. The proposed method classifies hand waving action accurately, but it gets 7% confused with the action waving while classifying the boxing action.

	Applauding	Waving	Boxing	Walking	Jogging	Running
Applauding	**1.00**	0.00	0.00	0.00	0.00	0.00
Waving	0.00	**1.00**	0.00	0.00	0.00	0.00
Boxing	0.00	0.07	**0.93**	0.00	0.00	0.00
Walking	0.00	0.00	0.00	**1.00**	0.00	0.00
Jogging	0.00	0.00	0.00	0.00	**0.95**	0.05
Running	0.00	0.00	0.00	0.00	0.03	**0.97**

Figure 3.7 Confusion Matrix for actions in KTH dataset

We compared the proposed methodology with the other state-of-the-art methods for KTH dataset in Table 3.1 Sadek et.al [125] uses the affine moment invariant of 3D action volume where they get the average accuracy of 93.3%. But to create the 3D volume of action there is a requirement of a sophisticated approach of background subtraction to detect the object. The action is represented with the combination of pose descriptors and negative space descriptor of each frame [47]. They get higher accuracy of 94.4% but still there is also required to have a very good background subtraction technique. Furthermore, [127] and [64] are getting very high accuracies of 95.8% and 96% respectively wherein [127] proposed a new feature Hanklets which are view-invariant features and makes the accuracy higher and [64] is traditional deep learning based methodology for action recognition that is presently used more popularly. The proposed methodology is outperforming these methodologies on KTH dataset as it is robust to view change, illumination change, and noise also. Table 3.1 shows that 98.2% action samples are correctly classified which is a very high accuracy as compared to other prominent methods of Table 3.1. In Table 3.2 comparison among the similar state-of-the-art methods is shown. The methodologies [135, 137-140] are based on the spatiotemporal features interest points. The disadvantage of these methodologies is that they do not show the relationship between the interest points and also the structural information of the clusters is missing. Although [135] retains the relationship among the spatiotemporal interest point, it is immune to noise. The proposed method shows the superior accuracy of 98.2% because it deals with these issues efficiently as discussed in the previous sections.

Table 3.1 Comparison among other state-of-the-art methods for KTH

Method	Year	EER	Accuracy
[30]	2015	5.8	94.2
[134]	2015	6.8	93.1
[125]	2012	6.7	93.3
[37]	2013	7.4	92.6
[47]	2014	5.6	94.4
[126]	2015	8.7	91.3
[127]	2012	4.2	95.8
[64]	2017	3.2	96.8
Proposed Method		1.8	98.2

Table 3.2 Comparison among similar state-of-the-art methods

Method	Year	EER	Accuracy
[137]	2014	5.8	95.0
[138]	2005	6.8	76.4
[139]	2011	6.7	84.1
[140]	2012	7.4	93.6
[141]	2013	7.5	92.5
[135]	2012	5.7	94.3
Proposed Method		1.8	98.2

3.4.2 Analysis and Discussion on Ballet Dataset

The Second dataset that we have used is Ballet. It is a typical dance dataset where different expressive dance steps. A single artist is performing the dance act. The dataset contains distinctive acts in Ballet dance such as hopping, jumping, standing still, left to right-hand opening, right to left-hand opening, leg swinging and turning right. This dataset has total of 44 numbers of videos. Our proposed methodology maintains the structural information of the clusters and thus classifies similar actions better than other methods as summarized in Table 3.3. Figure 3.8 shows the confusion matrix for actions in Ballet dataset.

As we can see from the matrix, actions such as standing still, leg swinging and turning right are accurately classified by the proposed method. Whereas hopping is 5% confused with jumping and jumping action is 8% confused with hopping. Similarly, Left to right-hand opening is 10% confused with right-hand opening and right to left-hand opening is 12% confused with the left to right-hand opening.

	Hopping	Jumping	Standing Still	LR Hand Opening	RL Hand Opening	Leg Swinging	Turning Right
Hopping	0.95	0.05	0.00	0.00	0.00	0.00	0.00
Jumping	0.08	0.92	0.00	0.00	0.00	0.00	0.00
Standing Still	0.00	0.00	1.00	0.00	0.00	0.00	0.00
LR Hand Opening	0.00	0.00	0.00	0.90	0.10	0.00	0.00
RL Hand Opening	0.00	0.00	0.00	0.12	0.88	0.00	0.00
Leg Swinging	0.00	0.00	0.00	0.00	0.00	1.00	0.00
Turning Right	0.00	0.00	0.00	0.00	0.00	0.00	1.00

Figure 3.8 Confusion Matrix for actions in Ballet dataset

Table 3.3 shows the comparison of the proposed method with some very efficient methods when they tested on Ballet datasets. Vishwakarma et.al [36] uses the silhouettes as the key poses that can express the actions. These silhouettes are represented into grids and cells. A fusion of classifier is used to recognize the actions. They achieved an accuracy of 94.2%. Their only disadvantage is that to represent poses the accurate silhouettes are required. Methods [45] and [135] represented the action with the set of spatiotemporal descriptors. Where [45] uses the dictionary learning approach and [135] uses the contextual constraint linear coding for action recognition. They showed 94.2% and 91.2% accuracy respectively. These methods are not robust to the data tempered by the noise. But as the proposed method is robust to noise that makes its accuracy comparable to these methods. The accuracy and the equal error rate of the proposed methodology are compared with other promising state-of-the-art methodologies in Table 3.3. The proposed methodology achieved 96.2% accuracy and 3.8% equal error rate.

Table 3.3 Illustration of EER and Accuracy for various approaches on Ballet dataset

Method	Year	EER	Accuracy
[128]	2008	48.7	51.3
[132]	2009	8.7	91.3
[50]	2014	9.2	90.8
[51]	2014	8.8	91.2
[135]	2012	5.5	94.5
[36]	2015	5.8	94.2
[45]	2012	8.8	91.2
Proposed Method		**3.8**	**96.2**

3.4.3 Analysis and Discussion on IXMAS dataset

The third dataset that we experimented on is IXMAS. There are five different cameras used for the videos in IXMAS dataset. For IXMAS dataset [135] five camera views are taken. We compared the proposed methodology with some latest prominent methodologies in Table 3.4. In Table 3.4 the methodologies [129] and [55] achieved the average accuracy of 88.3% and 88.5% on IXMAS dataset. [129] is based on the multiview human action recognition where several cameras are observing a scene and features from each camera are considered for recognition. It makes the system complex. In [55] a statistical translation framework is used where the cross view approach is applied. We find that we get the accuracy of our methodology for all five camera view is more than 90%. In the proposed method the camera view 3 shows the highest accuracy 92.4%. The overall accuracy computed by taking an average of all camera views is 91.2%.

There are different challenges included in this dataset that can reduce the recognition such as appearance change due to change in viewpoint and numbers of similar types of activities. Figure 3.9 depicts the overall confusion matrix where the action scratching head is only 5% confused with the action wave. Wave is 8% confused with the action scratching head. Getting up is 7% confused with the action picking up and picking up is 10% confused with the getting up. The confusion percentage is very less for these similar activities shows the dominance of the proposed methodology in the case of interclass similarity. Other actions are clearly classified with 100% classification.

Table 3.4 Accuracy for various approaches from 5 different cameras

Method	Year	C1	C2	C3	C4	C5	Overall Accuracy
[129]	2011	89.1	83.4	89.3	87.2	89.2	88.3
[133]	2010	84.2	85.2	84.1	81.5	82.6	83.4
[130]	2012	86.5	83.8	86.1	84.5	87.4	86.2
[131]	2013	91.3	85.7	89.3	90.2	86.5	88.5
[55]	2016	88.4	85.3	88.3	86.5	87.2	86.9
Proposed Method		90.8	90.6	92.4	91.2	90.6	91.2

	Check Watch	Cross Arms	Scratch Head	Get Up	Turn Around	Walk	Wave	Punch	Kick	Pick Up
Check Watch	1.00	0.00	0.00	0.00	0.00	0.00	0.00	0.00	0.00	0.00
Cross Arms	0.00	1.00	0.00	0.00	0.00	0.00	0.00	0.00	0.00	0.00
Scratch Head	0.00	0.00	0.95	0.00	0.00	0.00	0.05	0.00	0.00	0.00
Get Up	0.00	0.00	0.00	0.93	0.00	0.00	0.00	0.00	0.00	0.07
Turn Around	0.00	0.00	0.00	0.00	1.00	0.00	0.00	0.00	0.00	0.00
Walk	0.00	0.00	0.00	0.00	0.00	1.00	0.00	0.00	0.00	0.00
Wave	0.00	0.00	0.08	0.00	0.00	0.00	0.92	0.00	0.00	0.00
Punch	0.00	0.00	0.00	0.00	0.00	0.00	0.00	1.00	0.00	0.00
Kick	0.00	0.00	0.00	0.00	0.00	0.00	0.00	0.00	1.00	0.00
Pick Up	0.00	0.00	0.00	0.10	0.00	0.00	0.00	0.00	0.00	0.90

Figure 3.9 Confusion Matrix for actions in IXMAS dataset

To summarize, the overall performance of the proposed method shows that this is a very effective method for human action recognition. The three action datasets comprise of several challenges for human action recognition. The proposed method achieved 98.2% accuracy for KTH dataset, 96.2% accuracy for Ballet dataset and the average accuracy for all five camera views of the IXMAS datasets is 90.5%.

3.5 Chapter Conclusion

In this research paper, a new methodology that is the modified version of the bag-of-visual-word has been proposed. The limitation of the traditional codebook based method where the geometrical structural information of the clusters was missing has been overcome in the proposed methodology. Moreover, a new methodology for calculating the contextual distance among the points of the cluster has been used for action recognition. The proposed methodology has been validated on the challenging public dataset such as KTH, Ballet, and IXMAS where it shows its superiority as compared to other existing methods.

4. DESCRIPTOR BASED ON FUZZY LATTICES USING SCHRÖDINGER WAVE EQUATION

4.1 Methodology

In the case of the anomalous behavior of the crowd in a video, it is desired to monitor and analyze the sequence of activities happening in the video. To do so, a large number of frames of video are analyzed. While monitoring such large data, a lot of redundant information is found. To reduce redundant information, the paper has proposed dimension reduction approach based on fuzzy lattices using the Schrödinger wave equation. Block diagram of the proposed method is mentioned in Figure 4.1. The proposed method can be represented with the following points: (a) Representation of features in terms of lattices [142] with the help of optimized Gaussian fuzzy lattice function. (b) Expressing/measuring lattice features in terms of kinetic energy by the Schrödinger wave equation. (c) Classification of signal and decision-making process has been carried out to distinguish between normal and abnormal events by support vector machine.

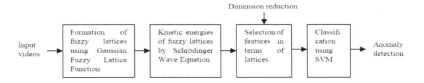

Figure 4.1 Block diagram of proposed method

4.1.1 Formation of Fuzzy Lattices

In every scene, there is nonlinearity in the image as the intensity value of pixels varies nonlinearly from one pixel to another. Therefore, this nonlinear relation between neighborhood pixels has been represented by performing nonlinear Gaussian fuzzy lattice function. This function has been given by the following equation and has been operated on each pixel:

$$f_{ij}(z) = e^{-\frac{(z-z_i)^j}{2\sigma_i^2}} \tag{4.1}$$

where z_i and σ_i are the mean and variance of the lattice function $f_{ij}(z)$ respectively, where z is the gray-level value of the pixel at coordinate positions x and y of the image. The values of i and j are taken from 1 to 4, because beyond this value, Gaussian will become an impulse. By doing this, the total sixteen number of Gaussian curves having different centers and width will be formed as lattice functions. These lattice functions are applied to each pixel. If the order of pixel connectivity is considered as 8, then every pixel in the image except the pixels at an edge of the image has 8 neighborhood pixels corresponding to eight directions, i.e., 45°, 90°, 135°, 180°, 225°, 270°, 315°, 360° respectively. The order of connected pixel is decided by the value of lattice functions. Each Gaussian lattice function-modeled pixel has been compared with the gray value of its neighborhood pixels within some degree of lattice function. Figure 4.2 explains the one-to-one connectivity among the pixels.

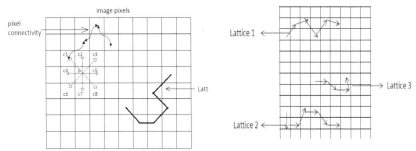

Figure 4.2   Connected Pixels   Figure 4.3   Fuzzy lattices

To find out the one-to-one connectivity, the first pixel is taken as test pixel and the neighboring pixels are found out. Now the connected neighboring pixel can be taken as the central pixel and the connected pixels are found out. This repetitive process yields a lattice of connected pixels and thus they form the sequence of connectivity. So the order of these features depends on the number of pixels connected in different directions. Figure 4.3 shows three lattices corresponding to three sequences of connected pixels. All the connected neighborhood pixels on the basis of equation (1) result in the formation of fuzzy lattices [77 and 90]. Fuzzy lattices can be represented by Equation 4.2

$$l_t = \cup_{ij} e^{-\frac{(z-z_{mj})^j}{2\sigma_{zi}^2}} \qquad \text{For t= 1, 2, 3... T} \qquad (4.2)$$

where l_t represents lattice which is the union of all connected pixels satisfying the relation established by Equation 4.1. In Equation 4.2, T is total number of lattices formed in an image representing a scene. Lattices behave like waves of the motion in the images and pixels are like wave particles. Therefore these lattices are expressed in the form of Schrödinger Wave Equation [80, 95] to find the change in the kinetic energy (K.E.) of the lattices which corresponds to any change in the dynamic features of an image or frame. Any change occurred in a scene due to motion of crowd can be represented by these fuzzy lattices. This change also produces the change in the kinetic energy of an image. This energy signal is a three dimensional signal where x and y co-ordinate of the pixel are responsible for change in K.E. due to change in position of x and y direction respectively. While grey level value of that pixel is responsible for change in K.E in z direction. To find x, y and z component of the K.E of these lattices, Gaussian lattice Functions of x; y and z are given by Equations 4.3, 4.4 and 4.5 respectively.

$$f_{ij}(x) = e^{-\frac{(x-x_{mi})^j}{2\sigma_{xi}^2}} \qquad (4.3)$$

$$f_{ij}(y) = e^{-\frac{(y-y_{mi})^j}{2\sigma_{yi}^2}} \qquad (4.4)$$

$$f_{ij}(z) = e^{-\frac{(z-z_{mi})^j}{2\sigma_{zi}^2}} \qquad (4.5)$$

x_{mi}, y_{mi}, and z_{mi} are means of x, y and z respectively and σ_{xi}, σ_{yi} are the variance of x and y co-ordinates of all pixels of a lattice and σ_{zi} is the variance of pixel value. Total Gaussian lattice function is given by Equation 4.6.

$$F = f(x).f(y).f(z) \qquad (4.6)$$

Suffixes i and j are ignored to reduce the complexity of the equation. Analysis of these lattices is done by Schrödinger Wave Equation by finding the Kinetic energy of lattices. These lattices can be treated as waves in different directions and pixels of lattice are like free moving particles, so there potential energy is zero. Change in Kinetic energy of lattices due to change in the position of pixels, is found out by solving Schrödinger Wave Equation (second order differential equation). From the solution of Schrödinger Wave Equation, kinetic energies of lattices formed in x, y and z direction are given by Equations 4.7, 4.8, and 4.9.

4. Descriptor Based on Fuzzy Lattices using Schrödinger Wave Equation

$$k.e_x = \frac{1}{f(x)} \frac{\partial^2 f(x)}{\partial x^2} \tag{4.7}$$

$$k.e_y = \frac{1}{f(y)} \frac{\partial^2 f(y)}{\partial y^2} \tag{4.8}$$

$$k.e_z = \frac{1}{f(z)} \frac{\partial^2 f(z)}{\partial z^2} \tag{4.9}$$

To find out the kinetic energy of the fuzzy lattices in the form of Equation 4.7, 4.8 and 4.9 we have taken the second order derivatives of fuzzy lattice functions expressed in Equation 4.3, 4.4 and 4.5. First order derivative of Equation 4.3 with respect to X direction is given by Equation 4.10 and the second order derivative is given by Equation 4.11.

$$\frac{\partial f_{ij}(x)}{\partial x} = \frac{-j}{2\sigma_{xi}^2} \{(x - x_{mi})^{j-1}\} e^{-\frac{(x - x_{mi})^j}{2\sigma_{xi}^2}} \tag{4.10}$$

$$\frac{\partial^2 f_{ij}(x)}{\partial x^2} = \frac{-j}{2\sigma_{xi}^2} \left\{ \frac{-j}{2\sigma_{xi}^2} (x - x_{mi})^{2j-2} + (j-1)(x - x_{mi})^{j-2} \right\} e^{-\frac{(x - x_{mi})^j}{2\sigma_{xi}^2}} \tag{4.11}$$

Similarly second order derivatives of Equation 4.4 and Equation 4.5 with respect to Y and Z directions are given by Equations 4.12 and 4.13 respectively.

$$\frac{\partial^2 f_{ij}(y)}{\partial y^2} = \frac{-j}{2\sigma_{yi}^2} \left\{ \frac{-j}{2\sigma_{yi}^2} (y - y_{mi})^{2j-2} + (j-1)(y - y_{mi})^{j-2} \right\} e^{-\frac{(y - y_{mi})^j}{2\sigma_{yi}^2}} \tag{4.12}$$

$$\frac{\partial^2 f_{ij}(z)}{\partial z^2} = \frac{-j}{2\sigma_{zi}^2} \left\{ \frac{-j}{2\sigma_{zi}^2} (z - z_{mi})^{2j-2} + (j-1)(z - z_{mi})^{j-2} \right\} e^{-\frac{(z - z_{mi})^j}{2\sigma_{zi}^2}} \tag{4.13}$$

To find out the kinetic energy of the fuzzy lattices in X, Y and Z directions, the second order derivatives from Equations 4.11, 4.12 and 4.13 are substituted in the Equations 4.6, 4.7 and 4.8 respectively to get Equations 4.14, 4.15 and 4.16.

$$k.e_x = \frac{-j}{2\sigma_{xi}^2} \left\{ \frac{-j}{2\sigma_{xi}^2} (x - x_{mi})^{2j-2} + (j-1)(x - x_{mi})^{j-2} \right\} \tag{4.14}$$

4. Descriptor Based on Fuzzy Lattices using Schrödinger Wave Equation | 2020

$$k.e_y = \frac{-j}{2\sigma_{yi}^2}\left\{\frac{-j}{2\sigma_{yi}^2}(y-y_{mi})^{2j-2} + (j-1)(y-y_{mi})^{j-2}\right\} \quad (4.15)$$

$$k.e_z = \frac{-j}{2\sigma_{zi}^2}\left\{\frac{-j}{2\sigma_{zi}^2}(z-z_{mi})^{2j-2} + (j-1)(z-z_{mi})^{j-2}\right\} \quad (4.16)$$

Kinetic energy of a pixel of fuzzy lattice in X and Y directions are given by Equation 4.14 and Equation 4.15 and Equation 4.16 represents its intensity value i.e. Z direction. If a lattice constitutes 'N' numbers of pixels, total kinetic energy of a lattice in X, Y and Z directions of a lattice is given by Equation 4.17, equation 4.18 and Equation 4.19.

$$total\ K.E_x\ of\ a\ lattice = \sum_{i=1}^{N} k.e_{xi} \quad (4.17)$$

$$total\ K.E_y\ of\ a\ lattice = \sum_{i=1}^{N} k.e_{yi} \quad (4.18)$$

$$total\ K.E_z\ of\ a\ lattice = \sum_{i=1}^{N} k.e_{zi} \quad (4.19)$$

Total Kinetic Energy of a lattice given in Equation 4.20 is calculated by adding energies in X, Y and Z directions from Equation 4.17, Equation 4.18 and Equation 4.19 respectively.

$$total\ K.E = \sqrt{(total\ K.E_x)^2 + (total\ K.E_y)^2 + (total\ K.E_z)^2} \quad (4.20)$$

Kinetic energy of a lattice calculated in Equation 5.20 may vary from one frame to another frame of a video due to motion.

4.1.2 Dimension Reduction and Classification

There are numbers of fuzzy lattices formed in the image due to motion. It increases the dimensionality of the feature. To deal with dimensionality problem we have adapted the following approach:

We have taken appropriate number of fuzzy lattices having maximum kinetic energies. Suppose X is a set representing all fuzzy lattices given by Equation 4.21

$L_d \varepsilon X$ where $L_d = 1,2,3...T$.   (4.21)

Number of fuzzy lattices selected decides the dimension of the feature. Lattices having maximum kinetic energy show the maximum change in a scene. Out of all those lattices suppose we are selecting n number of lattices having maximum kinetic energy where n will be the number of most prominent fuzzy lattices that are responsible for motion. The selection of n prominent fuzzy lattices is done on the basis of criteria mentioned in Equations 4.22-4.26 as follows: S_1 selects the lattice having maximum Kinetic energy. S_2 selects the lattice having highest Kinetic energy excluding S_1 whereas S_3 selects the lattice having highest Kinetic energy excluding S_1 and S_2. Similarly we calculate for n number of lattices S_n.

$$S_1 = max_{L_d} \{\text{Kinetic Energy}(L_1, L_2, L_3, ..., L_T)\} \quad (4.22)$$

$$S_2 = max_{L_d} \{ L_1, L_2, L_3, ..., L_T \setminus S_1 \} \quad (4.23)$$

$$S_3 = max_{L_d} \{ L_1, L_2, L_3, ..., L_T \setminus S_1 U\, S_2 \} \quad (4.24)$$

$$S_4 = max_{L_d} \{ L_1, L_2, L_3, ..., L_T \setminus S_1 U\, S_2 U S_3 \} \quad (4.25)$$

$$S_n = max_{L_d} \{L_1, L_2, L_3, ..., L_T\ S_1 U\, S_2 U S_3 U S_{n-1} \} \quad (4.26)$$

$S_1, S_2, ..., S_n$ are n fuzzy lattices having maximum kinetic energies from the set of fuzzy lattices X. Lattices having maximum kinetic energy (n) are used as the feature vectors. These features are fed to the Radial Basis Function based SVM classifier [93, 94] to classify the test video into abnormal or normal class.

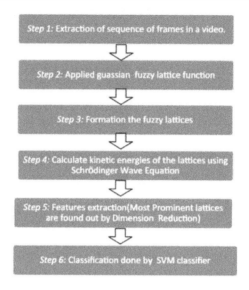

Figure 4.4 Process flow diagram for the proposed technique

The algorithm mentioned in Figure 4.4 explains the process flow of the proposed technique wherein step 1 extracts the sequence of frames in an input video. In step 2 pixels of a frame are modelled by Gaussian Fuzzy Lattices function. The connectivity of these pixels formed different fuzzy lattices in a frame in step 3. Step 4 calculates the kinetic energy of these lattices using Schrödinger Wave Equation. In step 5 dimensionality reduction is used to select the number of prominent lattices on the basis of their kinetic energies which is used as a feature vector. These feature vectors are fed to SVM classifier to detect the normal and abnormal class.

4.2 Datasets Used

The proposed method is implemented in MATLAB R2015a. and it is tested on system having hardware configuration processor Intel(R) Core(TM) i5 CPU @ 2.40 GHz ,RAM 8GB and 64-bit operating system. For real time analysis all the experimental results are validated on the CUDA enabled NVIDIA Titan Graphics Processing Unit using MATLAB 2015a. This method has been compared with the other state-of-the-art methods and applied for challenging and complex standard datasets UMN web dataset [96], UCSD [97] and UCF [98] for crowd activities.

UMN Dataset: The UMN dataset has three different scenarios where people are walking in different places and suddenly they started running to escape, this running activity is assumed to be anomaly in the video. Figure 4.5, Figure 4.6 and Figure 4.7 show examples of normal and abnormal frames of this dataset where first row of figure represents normal events in the scenes and second row represent abnormal events.

UCSD Dataset: UCSD dataset contains variety of anomalies occurring in crowd activities shown in Figure 4.8, Figure 4.9 and Figure 4.10. It uses mounted elevated stationary camera to monitor the pedestrian walk. Number of people walking in a scene varies from less crowded to more crowded scene. Normal event contains the walk of pedestrians and abnormal events contain people moving in random order, engaged in fights etc. In order to represent anomalies, dataset uses non pedestrian entities like bikers, skaters, small carts, wheel chairs. Another anomaly also introduced by adding anomalous pattern of walking of pedestrian. The videos of this dataset consist of two hundred frames in each video clip.

This dataset uses two scenarios.

Scenario 1: People are walking towards and away from camera. For training purpose 34 videos are used and 36 videos are used for testing purpose.

Scenario 2: People are walking parallel to camera. Sixteen videos are used for training and 12 videos are used for testing purpose.

UCF Dataset: UCF web dataset is developed by University of Florida. It is a promising dataset having real world crowd scenario shown in Figure 4.11, Figure 4.12 and Figure 4.13. This dataset contains 20 videos having normal activities of crowd which includes pedestrian walk. It contains 8 videos of abnormal activities which includes panic escapes, crowd fighting etc.

We have trained these datasets simultaneously to make a large dataset having different types of anomalies in the crowded place.

4.3 Parameter Sensitivity Test

Selection of the lattices is the most sensitive parameter of the proposed methodology. These lattices are selected on the basis of the maximum kinetic energies of the lattices based on Schrödinger Wave Equation. Movement in the scene changes the kinetic energy of the lattices in X, Y and Z direction. We have used the UCSD dataset for the parameter setting. We have tested it on different numbers 3,5,10 and 15. It can be observed that the proposed methodology is best suitable for 5. Although greater the number of lattices greater is the accuracy but for more than 5 lattices the overall accuracy varies from 2% to 3% only. Reason behind it that lattices having less kinetic energy contributes less in detection anomaly in crowd motion. More lattices increase the complexity and dimension in the methodology. That is why have selected 5 lattices having maximum kinetic energies. Similar setting is applied on the other dataset also.

4.4 Performance Evaluation Methodology

We used frame level performance evaluation methodology. For each frame in the video lattices having maximum kinetic energy are selected. We have compared proposed method with the other latest potential state of arts, qualitatively on frame level evaluation. We have selected two parameters, Area Under Cure (AUC) and Equal Error Rate (EER), which give relation between True Positive Rate (TPR) and False Positive Rate (FPR). Anomalous event are considered as "positive" and normal events are considered as "negative". TPR and FPR can be expressed by Equation 4.27 and Equation 4.28.

$$TPR = \frac{(True\ Positive)}{(True\ Positive + False\ Negative)} \qquad (4.27)$$

$$FPR = \frac{(False\ Positive)}{(False\ Positive + True\ Negative)} \qquad (4.28)$$

TPR are positive cases that are correctly classified and FPR are negatives that are incorrectly classified as positive.

EER basically gives the misclassification rate. EER can be defined by Equation 4.29 and Equation 4.30.

$$EER = 1 - Accuracy \qquad (4.29)$$

$$Accuracy = \frac{(TP+TN)}{(TP+TN+FP+FN)} \qquad (4.30)$$

where TP: True Positive; TN: True Negative; FP: False Positive; FN: False Negative. We used these parameters for comparing the proposed method to other state of arts. Performance evaluation discussed above is used to compare the proposed method with other efficient state-of-the-art different datasets.

4.4.1 Performance evaluation for UMN Dataset

Figure 4.5 shows the frame where a group of people is walking normally and Figure 4.6 shows the frame where people are running suddenly in different directions. Figure 4.7 shows that the ground truth for panic escape anomaly starts from frame no 95 to frame no 150. Proposed method detected the anomaly from frame no 100 to frame no 145. This means we are achieving high true positive from frame no 100 to frame no 145 and low false positive from frame no 95 to 99 and from frame no 146 to frame no 150. The reason of this low false positive is that we are taking the running average of the kinetic energies of 20 consecutive frames. This results in non detection of anomalies in the initial and the last 5 frames. If running average is not considered, a single high spike may also show an anomaly which may increase the rate of true negatives. Consideration of running average also helps to detect anomaly accurately by showing high spike for reasonable duration.

Figure 4.5   UMN dataset normal frame   Figure 4.6   UMN dataset abnormal frame

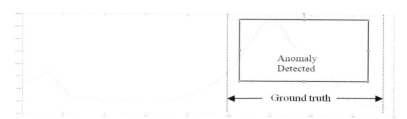

Figure 4.7   Result on UMN dataset (X axis represents frames and Y axis represents change in kinetic energy)

Equal error rate (EER) and Accuracy have been used for performance evaluation. Table 4.1 shows the comparison of our proposed method with other state of art methods on UMN dataset. Area under curve gives the area between true positive rates and false positive rate. Proposed method shows good EER of 1.6 as compared to other methods [78, 79, and 84]. AUC of almost all methods gives good result but AUC of our proposed method which is 99.8 shows the superiority over other existing methods. Table 4.2 shows the confusion matrix.

Table 4.1 EER and AUC as anomaly detection parameters on UMN dataset

Methods	Equal Error Rate	Accuracy
Dan Xu et. al [79]	4.8	99
Li et. al [78]	3.4	99.5
Mohammad Sabokou et. al [84]	2.5	99.6
Proposed Method	1.6	99.8

Table 4.2 Confusion Matrix of proposed method and other existing methods local dataset where N-normal and A-abnormal

Methods	N:Normal; A: Abnomal	N%	A%
Dan Xua et. al [79]	N	98.1	1.9
	A	6.2	93.8
Li et. al [78]	N	98.3	1.7
	A	5.9	94.1
Mohammad Sabokou et. al [84]	N	99.4	0.6
	A	4.8	95.2
Proposed Method	N	99.6	0.4
	A	2.4	97.6

4.4.2 Performance evaluation for UCSD Dataset

Figure 4.8 shows the frame where a group of pedestrian is walking normally and Figure 4.9 shows the frame where a non pedestrian (biker) enters the scene. Figure 4.10 shows the ground truth for non pedestrian anomaly starts from frame no 545 to frame no 643. Proposed method detected the true positive from frame no 550 to frame no 630 and false positive from frame no 545 to 549 and from frame no 631 to frame no 643. In UMN dataset example discussed above, the number of people is constant. Unlike UMN dataset, UCSD dataset has variation in the number of people in the scene. Some people are entering in the video whereas some are leaving, thus the curve in the result graph of this video is not as smooth as the curve of UMN dataset example. It has multiple changes in the spikes because kinetic energy changes every time a person enters or leaves the frame. But if the entry on new person gives sudden high spike in the result graph, it is detected as an anomaly.

Figure 4.8   UCSD dataset normal frame   Figure 4.9   UCSD dataset abnormal frame

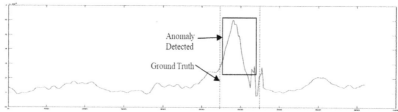

Figure 4.10   Result on UCSD dataset (X axis represents frames and Y axis represents change in kinetic energy)

Table 4.3 shows the comprison of our proposed method with other state of art methods on UCSD dataset. Most of the methods have been verified on this standard dataset. This dataset is simple, so all methods show good result. Proposed method shows better EER of 5.6 as compared to other methods. AUC of almost all methods gives good result but AUC of our proposed method that is 95.8 shows the superiority over other existing methods.

Confusion matrix shown in Table 4.4 represents that in proposed method only 2.8% normal class is confused with abnormal class and only 6.2% abnormal class is confused with the normal class.

Table 4.3 EER and AUC as anomaly detection parameters on UCSD dataset

Methods	Equal Error Rate	Area Under Curve
F. Yachuang et.al [86]	27.4	71.3
F. Zhijun et.al[99]	26.6	73.4
H. Mousavi et.al [100]	24.2	77.5
Dan Xu et. al [79]	18.5	80.4
Li et. al [78]	16.4	84.7
Mohammad Sabokou et. al [84]	14.2	87.2
Our proposed Method	5.6	95.8

Table 4.4 Confusion Matrix of proposed method and other existing methods local dataset where N-normal and A-abnormal

Methods	N:Normal; A: Abnomal	N%	A%
Dan Xua et. al [79]	N	78.4	21.6
	A	30.2	69.8
Li et. al [78]	N	81.4	18.6
	A	19.4	80.6
Mohammad Sabokou et. al [84]	N	86.6	13.4
	A	20.4	79.6
Our proposed Method	N	97.2	2.8
	A	6.2	93.8

4.4.3 Performance evaluation for UCF Dataset

Figure 4.11 shows people gathered and the police is warning the criminal and Figure 4.12 shows people running suddenly in all directions. Figure 4.13 shows that the ground truth for panic escape starts from frame no 74 to frame no 185. Proposed method detected anomaly from frame no 80 to frame no 180. Figure 4.6.c also shows that just like UCSD dataset, the result curve is not smooth because of variation in total number of people in the video.

Figure 4.11   UCF dataset normal frame   Figure 4.12   UCF dataset abnormal frame

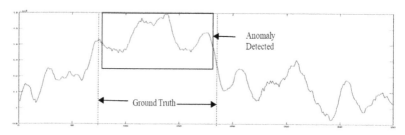

Figure 4.13   Result on UCF dataset (X axis represents frames and Y axis represents change in kinetic energy)

Comparison of EER of the proposed method with the other states of art methods shows the superiority of proposed method to over existing methods. Proposed method gives 12 percent EER evaluated in Table 4.5 which is very low as compared to other existing methods. Thus it outperforms other methods. Table 4.6 shows confusion matrix of our proposed method and other existing state of art methods. For our proposed method, only 2.6% normal class is confused with the abnormal class and 4.4% abnormal class are confused with the normal class. Thus, our proposed method clearly outperforms other existing state-of-the-art methods.

Table 4.5 Equal error rate for anomaly detection comparison for Local dataset

Methods	Equal Error Rate
F. Yachuang et.al [86]	22
F. Zhijun et.al[99]	21
H. Mousavi et.al [100]	35
Dan Xua et. al [79]	25
Li et. al [78]	20
Mohammad Sabokou et. al [84]	18
Our proposed Method	12

Table 4.6 Confusion Matrix of proposed method and other existing methods local dataset where N-normal and A-abnormal

Methods	N:Normal; A: Abnomal	N%	A%
F. Yachuang et.al [86]	N	85.8	14.2
	A	17.4	82.6
F. Zhijun et.al[99]	N	86.8	13.2
	A	12.2	87.8
H. Mousavi et.al [100]	N	79.6	20.4
	A	38.4	61.6
Dan Xua et. al [79]	N	75.6	24.4
	A	36.4	63.6
Li et. al [78]	N	86.4	13.6
	A	18.4	81.6
Mohammad Sabokou et. al [84]	N	94.7	5.3
	A	16.2	83.8
Our proposed Method	N	97.4	2.6
	A	4.4	95.6

4.4.4 Performance evaluation for YouTube video and New MOT 17 dataset

To validate the proposed method on other challenging environment we applied it on the MOT 17 [101] dataset and YouTube video [102]. YouTube video is shown in Figure 4.14. Video shows normal chores going on in the shop and suddenly a customer becoming violent with the shopkeeper. There are 2900 frames in the video. We used 1600 frames for the training and 1300 frames for testing purpose. Figure 4.16 shows that the ground truth for the video starts from frame no 470 to frame no 578. Proposed method detected true positive from frame no 480 to frame no 570. MOT 17 dataset is also analyzed for anomaly in the crowd activity. MOT 17-09-FRCNN video is used for training purpose. It contains the crowd Pedestrian Street captured from low angle stationary camera shown in Figure 4.15. It contains total 525 frame having resolution *1920 x 1080*. For testing purpose we used MOT 17-08-FRCNN which is also Pedestrian Street captured from stationary camera. Figure 4.17 shows that there is no anomaly detected in the scene. The graph has some high spike for very short duration which is not anomaly because it occurred due to the entry of set of new people in the scene. As people are disappearing from the scene it becomes very less crowded. This results the low value in the curve.

Figure 4.14   Youtube dataset   Figure 4.15   MOT 17 datase

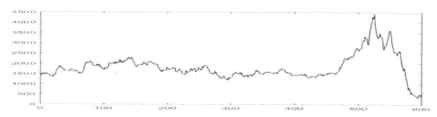

Figure 4.16   Result on Youtube video (X axis represents frames and Y axis represents change in kinetic energy)

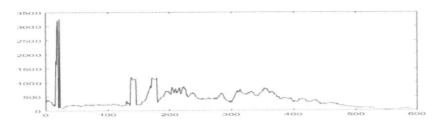

Figure 4.17 Result on MOT 17 dataset (X axis represents frames and Y axis represents change in kinetic energy)

4.5 Real-Time Applicability

We analyzed the performance of proposed method on UCSD datasets to find out the running time. We then compared our result with other state of the art methods. Frame resolution considered is *238 x 158*. For real time analysis we have used CUDA enabled Nvidia Titan GPU MATLAB 2015a having parallel computing tool box. We have calculated time duration for different steps used in the proposed method.

4.5.1 Run-time Analysis

Step to be performed according to proposed method and the time taken by proposed method in seconds.

1. Formation of Fuzzy lattices: 0.00132
2. Calculation of Kinetic Energy using Schrödinger Wave Equation: 0.00102
3. Selection of most prominent lattices: 0.0002
4. Classification: 0.0001

Total time for Anomaly detection: 0.00264

We are processing a single frame in 0.00264 sec. Proposed method is using 378 frames per second for processing which makes it real time applicable. Table 4.7 shows comparison in the speed of the proposed method with other state of art methods. Proposed method is giving far superior result as compared to the other state of art methods. Even if compared with recent methods based on CNN [85, 87 and 88] with fast running time, our method shows dominance. FCN [89] based method that is extended version of CNN is achieving very good real time applicability (340 fps). But proposed method is performing better as it is using 378 fps. One more reason this method is real time applicable is its low dimensionality. Table 4.8 shows the comparison between different dimension reduction techniques. Proposed method uses only 5 dimensions.

Table 4.7 Run-time analysis of different methods

Methods	Run-Time (seconds)
Dan Xua et al [79]	Offline
Li et al[78]	1.34
Mohammad Sabokou et. al [84]	0.04
Karpathy et al [87]	0.007
Meng et al [88]	0.004
Sabokrou et al [85]	0.0027
Proposed Method	0.00264

Table 4.8 Comparison of proposed method with other dimension reduction technique

Methods	Dimensions	Recognition Accuracy (%)
PCA	42	80.5
LLE	32	86.4
Laplacian	15	78.4
Isomap	10	88
Proposed Method	5	96.6

4.6 Chapter Conclusion

The proposed method is a novel technique for anomaly detection in a video, where the concept of mathematics, physics, and computer science along with signal processing has been combined to model a detector for anomalies. It is capable of determining anomalies in real-time because it considerably reduces the dimensionality of a feature. Nonlinear Gaussian membership functions (GMF) have been applied to video for finding the dynamic features in terms of fuzzy lattices and their corresponding tensors. This method used the concept of the Schrödinger wave equation for modeling these features. Energy signal, obtained after applying the Schrödinger wave equation is classified into normal and abnormal events. To validate the proposed method, standard datasets such as UCF web dataset, UCSD dataset, and UMN dataset are used. The experimental result shows the superiority of our proposed method over other state-of-the-art methods for anomaly detection. The proposed method has also been applied to YouTube video and MOT17 dataset which gives good results.

5. CONCLUSION AND FUTURE SCOPE

Conclusion

We proposed new methodologies that can deal the issue of global and local features descriptors. We also proposed new methodology that can detect anomaly in the crowded area. We also proposed methodology that used reduced dimensional feature and real time applicable. We also modified traditional Bag-of-Visual-word based methodology to give the information of structure of the cluster.

- We proposed new local feature descriptor based on Finite Element Analysis (FEA). These feature vectors are represented in terms of stiffness matrices of the silhouette of the action video. They extract both shape and motion information of the action video. Hybrid feature descriptor is being used to represent shape and motion features but this descriptor increases the feature dimensions significantly. The proposed feature descriptor is capable of representing shape and motion information. The run time of the proposed methodology is considerably less. The feature vectors extracted from the proposed method are given to the RBF-SVM classifier. The proposed methodology gives better result as compared to other existing methodologies when applied on the challenging and standard datasets: Weizmann, KTH, Ballet and IXMAS.

- We proposed modified bag-of-visual-word based methodology. The traditional bag-of visual word based methodology is suitable to deal the difficult challenges like background cluttering, variation in view point and occlusion. But the traditional bag-of-visual-word based methodologies do not retain the structural information of the clusters. In the proposed methodology the contextual distance among the points of the cluster is calculated to retain the structural information of the cluster. This methodology clearly increased the recognition rate even for the similar types of actions. The proposed method gave better results when compared to other existing methodologies.

- We proposed a novel technique for anomaly detection in the crowd activities. We proposed new feature descriptor based on fuzzy lattices using Schrödinger Wave Equation. This method is applicable in real time environment because it uses reduced number feature dimension. The fuzzy lattices based features are extracted from nonlinear Gaussian membership function. These fuzzy lattices are described by the kinetic energy using Schrödinger wave equation. The feature vector is formed on the basis of the values of the kinetic energy of the fuzzy lattices. The maximum kinetic energy shows maximum change due to crowd motion. We compared the proposed methodology with other existing methodologies and we get better results. We have used variety of standard datasets such as UMN, UCSD, UCF, MOT 17 and Youtube datasets to validate the proposed methodology.

Future Research Scope and Application

We proposed a new local feature descriptor based on fuzzy lattices and the famous Schrödinger equation. This can be used to detect motion in a video efficiently. We also proposed new feature descriptors based on Finite Element Analysis. This feature is capable of capturing motion as well as shape information of the human body. In future they can be used in the field of biometric for gait recognition, gesture recognition and in the smart class rooms also. Cameras installed for the security is very crucial in today"s scenario. Bag-of-visual-word based methods are robust to the challenges like view variation and background cluttering. These methods have misclassified due to its incapability of retaining structural information of the cluster. We modified the existing Bag-of-visual-word method through retaining the structural information of the cluster makes it more reliable. In future this method can be implemented to find out the anomalies in the video because it is invariant to view and occlusion. In future, we may develop a methodology that is capable to deal the problem of occlusion, especially self occlusion.

Ingram Content Group UK Ltd.
Milton Keynes UK
UKHW021431070623
423003UK00008B/53